THE
UFO
FILES

LOOKING FOR THE ALIENS

JENNY RANDLES AND **PETER HOUGH**

D1082631

BLANDFORD

First published in the UK 1999 by Blandford
A Member of the Orion Publishing Group

Cassell & Co.
Wellington House
125 Strand
London WC2R 0BB

Previously published by Blandford as *Looking for the Aliens*

Distributed in the United States by Sterling Publishing Co., Inc.,
387Park Avenue South, New York, NY 10016-8810

A Cataloguing-in-Publication Data entry for this title is available and may be
obtained from the British Library

ISBN 0-7137-2800-0

Printed in Great Britain by Cox & Wyman Ltd, Reading, Berks.

Contents

The Consequences

Introduction

Aliens – a word which triggers a response in every one of us. We might use it to describe comic-book flying saucers and little green men, *Star Trek*, NASA's search for extraterrestrial life, or some abomination which has leapt out from the pages of a novel by Stephen King.

All these stem from a deep-rooted, perhaps intrinsic belief in us that there is something out there, in the interminable blackness of outer space – another dimension where the terms 'fantasy' and 'reality' cease to have any clear-cut meaning.

To many space scientists, the word 'aliens' is used to describe the creatures living somewhere else in the universe, separated from us by unimaginable distances. One day we might receive their radio transmissions, just as they might receive ours – perhaps thousands of years after we have died out. To the fiction writer, aliens are often a metaphor for some cold, calculating horror within ourselves. To the modern physicist, engrossed in the quantum physics revolution, the cosmos becomes a ghost universe, and the concept of alien worlds takes on new proportions. And for some people, alien beings have been here all along, manipulating and chastising the human race from the wings.

In this book we have tried to bring together all the different concepts of aliens. This has never been done before. As an intelligent species, we tend to compartmentalize things, starting off with what is 'reality' and what is 'fantasy'. The trouble with this game of logic is that someone keeps moving the goal posts and changing the rules. Books have been written specifically on the genres of science fiction and horror, SETI, quantum physics, 'channellers' and UFOs. What we have done is taken all of these fields and thrown them *en masse* into a huge melting-pot. When it comes to aliens, we believe the rule book should be discarded.

'They' – the aliens – are, by definition, separate from us: yet they are also within us, part of our inquisitive nature that is always wondering if there is life 'out there'.

We have solicited the views of fiction writers such as Bob Shaw, Ramsey Campbell, David Langford, John Grant and Arthur C. Clarke in our search. The SETI institute in California has furnished us with much information on NASA's own search for aliens, and the views of physicists are aired. We explore some key UFO incidents which seem to indicate that something other than misperception is at work, and which are regarded by some authorities as proof of alien visitors. And on the outer edge of alien contact, the claims of people allegedly on the receiving-end of psychic messages from extraterrestrials are discussed. We have asked what the ramifications are for society and religion if one day the world should wake up to hear the awesome news: The aliens are here.

Inevitably, we could not resist the temptation to 'compartmentalize' a little. Consequently, this book consists of five sections: The Dream, The Belief, The Search, The Evidence, and The Consequences. Yet in a way, these headings are arbitrary. What is a dream for one person is reality for another. What one individual will cite as evidence, another will ignore. We merely tell the story of the human relationship with an inner belief, hope and often fear, in alien existence.

Nevertheless, whether this existence is dream or reality, we hope our book will provide fresh food for thought, and prepare you for the inevitable day.

The search is on – we are looking for the aliens.

Peter Hough
Jenny Randles

The Dream

1
Who Goes There?

Our reality, objective and subjective, is structured around our ability to imagine. This abstract function of the right hemisphere of the brain is often loudly and crudely derided. The charge 'he has too much imagination' is a label for feeble mindedness which is applied to someone who is 'out of touch' with 'reality'. Yet imagination is the catalyst for all human endeavour, whether it is good or bad, significant or insignificant. Like it or not, all of us fantasize every day of our lives. A job interview does not begin at the point of entry into the interview room. It begins long before that – in the imagination. Before we reach the threshold of that oak-panelled door, we have already seen beyond it, met the signatory of the invitation letter, shaken their hand and told them our somewhat embellished career history. The reality may differ markedly from what we imagined, but we needed the fantasy to prepare us for what was to come, to anticipate questions and prepare answers, just in case . . .

Many inventions, such as the helicopter, the submarine, the electronic computer and the humble sewing-machine needle, owe their genesis to this much maligned faculty – an unfettered wild thing which goes against the grain of 'common sense' and 'logic': an *alien* function. Yet it is out of this chaos that order triumphs.

In our search for the aliens, it makes sense, therefore, that we should start with the dream, and examine how some writers have harnessed the imagination to tell us what contact with extraterrestrials might really be like. Just in case . . .

Without doubt, one of the most powerful literary statements on this subject is H. G. Wells's *War of the Worlds,* first published in book form in 1898. This is a cataclysmic tale of invasion by an alien race which shows no quarter. The Martians, sinister octopus-like beings hidden from view in the turrets of monstrous tripodal machines, search out and destroy everything in sight, with a mechanical callousness. There is no alien contact here. The aliens have come to do a job, and they do it as dispassionately as building developers bulldoze a piece of the countryside in readiness for their houses. Victorian armaments are useless against these invaders. It is the earth itself which releases a deadly cocktail of 'antibodies' – microbes – which finally destroys the invaders.

This cover illustration by Frank R. Paul, from an August 1927 edition of the pulp magazine *Amazing Stories*, depicts H.G. Wells's *War of The Worlds*, (*Frank R. Paul*)

This powerful vision of alien invasion spawned a film, a radio play and a rock album. The broadcast, masterminded by a young Orson Welles in 1938, convinced many people in the United States that an invasion really was taking place and caused widespread panic.

The cold-blooded genocide of the human race by Wells's Martians should not necessarily be seen in terms of sadism. UFO abductees describe a similar attitude of their abductors as the aliens coldly go about the sometimes painful examination of their captives' minds and bodies. Whitley Strieber, in his book *Communion*, said that the beings who experimented on him were not purposely cruel but acted as if they were just doing a job. He likened them to insects, possessing some sort of hive mind. Former police officer and UFO abductee Philip Spencer, reported a similar thing when he said of his abductors: 'They worked like bees.'

Wells really gets to grips with the aliens in *The First Men in the Moon*, published in 1901. Here, contact takes place when two Englishmen, Mr Bedford and Mr Cavor, build a spaceship, powered by 'cavorite', an anti-gravity material, and travel to the moon, There they become lost in an underground labyrinth. This is how Mr Bedford describes the 'Selenites': 'Scarcely five feet high . . . having much of the quality of a complicated insect . . . there was no nose, and the thing had dull bulging eyes at the side. There were no ears.'

Olaf Stapledon as depicted by Barry Robson. Stapledon was author of *The Star Maker*, published in 1937. In it, a man involuntarily leaves the shackles of his body and starts a journey around the universe, collecting other minds. Eventually, this new mind-swarm meets the ultimate 'alien'. (*Barry Robson*)

There were tentacles on the creatures, too, but, interestingly, the insect motif is the dominant image. In fact the Selenites *are* insects. Unhindered by the Earth's greater gravity, this ant-like species has developed on the moon not only physically, but also intellectually. Many American abductees, among them Strieber, describe entities similar to Wells's Selenites, with large domed heads, large black insect-like eyes and stick-thin arms and legs.

However, extraterrestrials come in all shapes and sizes, limited only by the skill and imagination of the science fiction writer. In 1937, a classic novel of alien types, *The Star Maker*, was published by Olaf Stapledon. The narrator leaves his body and wanders the universe, visiting every type of humanoid race imaginable. There are extraterrestrials with various skin pigmentations and taste glands in palms, feet and even in genitals! Whale-like nautiloids, flying men, symbionts and plantmen also appear. This disembodied narrator collects one mind from each race to continue with him on the journey. When every alien civilization has been visited in this way, and the minds combined, only then is the existence of the Star Maker understood.

Perhaps one of the best pieces of science fiction, which digs at the very roots of our fear of intelligent alien life forms, is the classic novella *Who Goes There?* by John W. Campbell, published in 1938. (Campbell was also an editor responsible for nurturing the prestigious talents of such authors as A. E. van Vogt, Robert A. Heinlein, Isaac Asimov and Theodore Sturgeon.)

The action of *Who Goes There?* takes place in an Antarctic camp staffed by scientists. An expedition from the camp discover a crashed spaceship which has been buried in the ice for 20 million years. Near the ship, encased in ice, is an 'animal'. It is carried back and scientists argue about allowing it to thaw. This is what physicist Vance Norris has to say to an enthusiastic colleague:

'Damn it, Blair, let them see the monstrosity you are petting over there. Let them see the foul thing and decide for themselves whether they want that thing thawed out in this camp.

'Unwrap it, Blair. It may have a different chemistry. I don't know what else it has, but I know it has something I don't want. If you can judge by the look on its face – it isn't human so maybe you can't – it was annoyed when it froze. Annoyed, in fact, is just about as close an approximation of the way it felt as crazy, mad, insane hatred. Neither one touches the subject.

'They haven't seen those three red eyes, and that blue hair like crawling worms. Crawling – damm, it's crawling there in the ice right now!'

Real creepy stuff, and that's *before* the thing comes alive! When it does, it disappears – or does it? Actually, it does not, for the alien has the perfect survival mechanism. It can imitate any living thing. Hence the title. Which of the crew is the thing?

In 1951, *Who Goes There?* was filmed as *The Thing From Another World*. This was a dreadful film, an unfaithful adaptation with poor visual effects, and compared to Campbell's masterly vision, it was immature. Another attempt was made in 1982, when John Carpenter directed *The Thing*. It's script and special effects were stunning. The only flaw came at the end, when Carpenter, for unknown reasons, decided to abandon Campbell's mind-blowing denouement.

In the genre of science fiction, it is very difficult to create really *alien* extraterrestrials, for reasons which Bob Shaw explains later. But Lord Lytton is one of the few writers who comes close to it in *The Coming Race*, published in 1874. Here the aliens are supermen who live deep beneath the earth's crust. They have great physical and mental powers, but the real difference between us and them is the structure of their society. It lacks crime, government and war. Everything is based on common consent – and that is alien!

Out of the dozens of more recent novels and short stories which deal with aliens, several stand head and shoulders above the rest. Two of them are Larry Niven and Jerry Pournelle's *Footfall* and Greg Bear's *The Forge of God*. The latter, especially, keeps the aliens 'alien'. The reader, like the human characters in the book, never finds out exactly what the aliens are. Bear successfully maintains a juggling act which allows us to learn enough but not too much – and the mystery and the wonder are preserved.

Fear is one of the key emotions in the alien equation. It is utilized most effectively when the normal is used as a mask for the abnormal. John Wyndham plays on our fear in *The Midwich Cuckoos*, effectively filmed as *Village of the Damned*. In it, an entire village is put to sleep, during which a number of women are impregnated. The little horrors which emerge seem sweet enough until they start using their alien telekinetic abilities.

2
Aliens – Haunters of the Dark?

Clearly, the lingering local rumours had not lied. This place had once been the seat of an evil older than mankind and wider than the known universe . . .

Before he realised it, he was looking at the stone again, and letting its curious influence call up a nebulous pageantry in his mind. He saw processions of robed, hooded figures whose outlines were not human, and looked on endless leagues of desert lined with carved, sky-reaching monoliths. He saw towers and walls in nighted depths under the sea, and vortices of space where wisps of black mist floated before thin shimmerings of cold purple haze. And beyond all else he glimpsed an infinite gulf of darkness, where solid and semisolid forms were known only by their windy stirrings, and cloudy patterns of force seemed to superimpose order on chaos and hold forth a key to all the paradoxes and arcana of the worlds we know.

(*Haunter of the Dark*)

Howard Phillips Lovecraft, Edgar Allan Poe, Ambrose Bierce and Nathaniel Hawthorne were the unconscious architects of the horror tradition in America. There were very many British writers of note who also added richly to the body of this literature – Algernon Blackwood, M. R. James, Arthur Machen, John Collier and L. P. Hartley – but it is Lovecraft who is most often referred to by modern word-weavers of the genre. More than the others, Lovecraft used the entire universe as his canvas, and the barely perceptible creatures within it to make us aware of its dark complexities. The opening quote from *Haunter of the Dark* illustrates the way in which the alien entity became fundamental to horror fiction. Lovecraft's extraterrestrials, the most notorious being 'Cthulhu', became more than mere story characters, but took on a life of their own, appearing in the writings of others.

Born in 1890, of British ancestry, Lovecraft spent most of his life in Providence, Rhode Island, before dying prematurely in 1937, just 20 years after he began producing the strange literary work for which he is remembered. Yet, like a lot of great writers, public acclaim did not emerge until after death. Now Lovecraft has a global following, and his work regularly appears in anthologies, as well as being held up as a shining example by modern-day authors of the literature of inner and outer space. One such author who answered the call of Cthulhu was British writer, Ramsey Campbell.

Top British horror writer, Ramsey Campbell, pictured here with son Matty, in the library of their home in Wallasey, Merseyside. Campbell believes that the alien concept is a metaphor for something much deeper, which affects us all. (*Peter A. Hough*)

Campbell is large, loud and jovial – not at all like the frail, reclusive Lovecraft. If there is any similarity, then it is a tenuous one of background. As a boy, Campbell shared a house with a father, whom he never saw 'face to face' for 20 years, and a schizophrenic mother who was convinved that 'they' had a conspiracy against her. Until her death, Lovecraft lived alone with his mother, who frequently reported the sighting of strange creatures. That is where the similarity ends: while Lovecraft was a social failure whose marriage lasted barely two years, Campbell is a welcome addition to any dinner party and is happily married with two splendid children.

Born and bred in Liverpool, Campbell took the fear and supernatural entities of Gothic horror and relocated them in the busy streets and dark, draughty houses of his childhood — a call taken up since by another writer from Liverpool, Clive Barker. Far from diminishing in terror, Campbell's fiction is more unsettling to the faint-hearted because now these creatures of the dark are literally brought home to us.

Campbell's first collection of short stories was published when he was just 18 years old through the guidance and encouragement of Lovecraft's old protégé, August Derleth. Since then he has written well over 200 short stories and 10 novels, including his recent, *Midnight Sun*. Fans and peers alike have rewarded his efforts with several awards for 'best short story', and August Derleth Awards for his novels *The Parasite, Incarnate, The Hungry Moon* and *The Influence*.

Ramsey Campbell crossed over the River Mersey to Wallasey about seven years ago, and now lives in a semi-detached Victorian mansion bearing more than a passing resemblance to the Bates's house in Hitchcock's *Psycho*. From his study on the third floor, the legendary (and polluted) river is visible over distant roof-tops. The Mersey perhaps epitomizes the sweet and sour excesses of its famous son's fiction.

After browsing through the library, which boasts in excess of 2,000 books and magazines, we followed Campbell's disembodied voice down the stairwell to the 'east wing', where coffee was being poured. The interview began:

The alien theme does not just emanate through the science fiction genre, but is alive and kicking within the literature of horror too, albeit often in disguise. Do you agree?

'Yes, I do indeed! There's been a process through the twentieth century emerging at the point of Lovecraft, where horror became science fiction and vice versa. There's a strong argument that Mary Shelley straddled that gap too. One of the crucial things about *Frankenstein* is that the monster becomes more and more identifiable with its creator. When the moment arrives where Frankenstein comes face to face with the monster, we realize it is his alter ego. One of the central issues of horror, is not so much the monster – the alien – *per se*, but the idea *that it is something about ourselves*. I believe it is a crucial element within us which has not yet been identified.'

Didn't Lovecraft see it more in conventional terms? The Cthulhu mythos is a body of literature of a race of banished aliens who occasionally break through and confront individuals in their attempt to regain control of our reality.

'There is an oft-quoted piece of text which says: "All my stories, unconnected as they may be, are based on the fundamental lore or legend that this world was inhabited at one time by another race who, in practising black magic, lost their foothold and were expelled, yet live on outside ever ready to take possession of this earth again."

'I think it is pretty well established that that wasn't Lovecraft's own statement, but something created by August Derleth – who I must say I have nothing but respect and great gratitude for. He has promoted me as a writer. But I think there was a problem here because Derleth was a Christian, and Lovecraft was an atheist.'

Indeed, that statement appeared in an introduction written by Derleth for *The Haunter of the Dark – Omnibus 3*, first appearing in 1950, currently a paperback reprint. He does attribute it to Lovecraft and then goes on to say: 'The similarity of this pattern to the Christian mythos, particularly in relation to the expulsion of Satan and the powers of evil from Eden, is evident.'

Campbell:

'Lovecraft was an atheist, and a fairly pessimistic atheist at that. But he came up with the concept of a notion of alienage which was a metaphor for something barely glimpsed, something larger than the text – a cosmic horror with a visionary quality. Algernon Blackwood achieved this with his outstanding novella *The Willows* and to a lesser extent with *The Wendigo* – Robert Aickman's favourite Blackwood story.'

The 'wendigo' is an entity which lives in the Canadian forests, and it has the ability to take on the form of any human being. In this way it can get close to the hunters and trappers before wreaking out revenge. In this sense the story is similar to John Campbell's *Who Goes There?* discussed in Chapter 1. But the wendigo is a product of the forest, so in this case it is the human interlopers who are the 'aliens'. When it takes on the form of a man, the only way it can be detected it by its misshapen feet.

'The main difference between Blackwood and Lovecraft was that Blackwood seemed to think that the other lacked a spiritual dimension. But I think that is what precisely gives Lovecraft his power. There's no spirituality, but there is a compensating awe at what may lie beyond the blue sky we all look out on every day. That sky is comforting during the day, but after the sinking of the sun, it is replaced by something terrifying and totally alien.

'The Lovecraft mythos conveys, in a metaphor, those huge but indifferent forces from out there which occasionally seep into human activities.'

The idea that extraterrestrials (or even *ultra*terrestrials) might be indifferent towards the human species spills over into UFO abduction accounts. Many alleged abductees describe how the entities treat them with the cold precision of laboratory technicians experimenting with rats. While we can cope quite happily with aliens who want to kill and enslave us, aliens who treat us like farm animals are another matter entirely. It illustrates a gulf which is too wide: it knocks us off our pedestal, crushes the ego and brings it home to us that we are not the be-all and end-all of creation.

'Yes, exactly! So we look out at the night sky and think: this is gigantic. But that is nothing. Beyond the farthest reaches of as far as we can see – that only takes us to the edge of the beginning. This unchartable, unknowable vastness goes on and on and on . . . To *it* we are nothing.

'Lovecraft understood this. The more Lovecraft wrote, the more suggestive he became, and his aliens became less monstrous and more a vehicle for perceiving further. Michael Moorcock – a writer I usually have a lot of respect for – commented not too long ago that he thought Lovecraft had retreated to a position of permanent defensiveness. It seems to me that, on the whole, he did the reverse; from the monstrous he went to the cosmic. An awesomeness comes through in tales like *At the Mountains of Madness* and *The Shadow Out of Time*. He had gone past the need to be repulsive. In fact, the creatures in *At the Mountains* – although very alien – are explicitly identified with humanity. I think often the aliens are a metaphor for something which is a part of us.'

The Shadow Out of Time concerns an ancient 'Great Race' of aliens who conquered the earth in the Mesozoic period. The aliens had the ability to transmigrate and took possession of a species of evolutionary dead-ends which were part-vegetable, part-animal.

At that time the earth was already in the grips of a predatory race of beings, which the newcomers imprisoned deep underground, then moved into their great stone cities. Having mastered time and space travel, the Great Race sends out travellers to possess the bodies of other beings to learn and to study the cultures of other societies. One ousted being, an American professor called Peaslee, finds himself a prisoner in an alien's body back in the Mesozoic era. On his return to his own body, he suffers amnesia, punctuated by strange dreams which gradually fill in the period of missing time.

But it is the predatory beings which occupy Peaslee's mind. These he discovers are still living in huge underground caverns beneath the Australian continent, long after the Great Race has gone. Lovecraft never lets us see them, but hints at what they might be like:

They were only partly material – as we understand matter – and their type of consciousness and media of perception differed widely from those of terrestrial organisms. For example, their senses did not include that of sight; their mental world being a strange, nonvisual pattern of impressions.

At no time was I able to gain a clear hint of what they looked like. There were veiled suggestions of a monstrous plasticity, and of temporary lapses of visibility, while other fragmentary whispers referred to their control and military use of great winds. Singular whistling noises, and colossal footprints made up of five circular toe marks, seemed also to be associated with them.

Down the centuries there have been many legends of beings who live in deep underground caverns. Some early adherents of flying saucers believed they came from the centre of the earth, via holes at the poles. At one point a satellite photograph was presented which allegedly showed such a hole, but this turned out to have a prosaic explanation.

However, in the literature of horror, the aliens are more likely to be interdimensional than just plain extraterrestrial. One of Campbell's own tales *The Voice of the Beach*, is a good illustration. The beach in question is a place where this world, and another, merge. It seems alive, an unearthly glow pervades it at night; bushes rustle without a wind; and when a man disappears, his likeness appears in the sand. *Their* world becomes ours in a perversion of reality.

Is horror and science fiction a perversion of a hitherto-denied reality, or are encounters with alien beings a crude and distorted reflection of the writings of the perfect dreamers — our master story tellers? What of the thousands of unexplained UFO reports, the detailed alien abduction cases which continue to baffle mental health specialists? What of the incubi, succubi and other demonic entities which both the Christian Church and occultists believe in? What of the most horrific and evil alien of all — Satan? Are we merely talking about a section of the population who misperceive fantasy as reality? Or do the awesome powers written about by the likes of Ramsey Campbell actually exist?

'The classic take-over by science fiction of the Christian myth must be Arthur C. Clarke's *Childhood's End*. At the close of the first section of the novel, we are faced with an entity sporting horns and a tail. Yet this clearly does not represent some terrible evil, but the next step in evolution. I think that Clarke was on to something. The human mind manufactures metaphors in an attempt to grasp some niggling feeling, some fragmentary glimpse of the unknown.

'That unknown can be something very beneficial, yet also seductive and dangerous. But that is the crux of all the striving for scientific knowledge — a journey out on to the fringes of the unknown, grasping that shadowy *something*, struggling to make sense of it.'

Interestingly, some ufologists, mainly British, have toyed with the idea that the UFO entities, and subsequent abduction incidents encountered by people, might be a 'cover story' — in a sense a metaphor — for something else totally beyond human understanding.

'Yes, and if we are to breach this metaphor — as many have done throughout human history in other areas of science — we must make The Bargain. The legendary Faust sold his soul to the Devil in exchange for greater knowledge, and we have to sacrifice something too. Goethe, Mary Shelley and Robert Aickman thought that bargain a loss. When you've lost the spiritual comfort that religion gives, you've got to seek it elsewhere. I find it through listening to great music. I also find it in great horror fiction.'

Arguably the greatest living American horror fiction writer is Stephen King. Certainly on output alone he deserves that award; blockbuster

An evocative illustration for Ramsey Campbell's *The Voice of The Beach* by David Lloyd on the cover of the British magazine *Fantasy Tales*. The beach in question is in flux between our world and another, utterly alien environment. (*David Lloyd*)

THE VOICE OF THE BEACH
BY RAMSEY CAMPBELL
A WITCH FOR ALL SEASONS
by Manly Wade Wellman

novel follows blockbuster novel: *Carrie, Salem's Lot, The Shining, IT, Misery* – to name a few, plus his collections of shorter tales. Often the plots are basic and unoriginal, but King is a true writer. His characters walk off the page and his prose pushes the darkness aside just enough for us to see our way. When the blackness becomes too much, a pinch of humour lightens things up. His books are best sellers because they deserve to be.

With *The Tommy-Knockers*, King broke new ground. Here was a horror novel that dealt with a science-fiction premise: What would be the consequences of someone digging up an alien spaceship which had been buried beneath the earth millions of years before? Of course we have met this scenario already in *Who Goes There?* and a thousand short stories written since; but that is where the similarity ends. King's flying saucer is not buried in remote Antarctica but on land belonging to Bobbi Anderson, Maine, USA.

After Bobbi trips over a piece of metal jutting out of the ground during a walk with her dog, it becomes an obsession to discover what is buried. Is it a car? An old abandoned refrigerator? She solicits the help of an alcoholic poet called Jim Gardner. But the obsession is not natural, but induced by whatever has been lying dormant in the spaceship down all those countless centuries. Gardner is immune to the mind-controlling effects of the ship because of a metal plate in his head, but he is so drunk most of the time anyway that it makes little difference. Under instructions from the ship, revolutionary earth-moving equipment is manufactured by Bobbi and a number of other affected villagers to release the disc or spaceship — which is thousands of metres across.

But the villagers become the aliens as physiological and mental changes take hold of them. Is this a further example of 'the aliens are something in us' theme? Certainly that is what Ramsey Campbell seems to think.

'Here we are, sitting opposite one another on these chairs, we look out of the window and as a writer I can make you see strange things. There is a dissatisfaction with consensus reality – what Arthur Machen would call 'the veil', and Lovecraft would identify as the banal and the mundane. Is Ufology a striving for the larger-than-can-be-perceived-than-can-be-stated?'

Campbell sees quite clearly that the key to our search for the aliens lies with the imagination.

'We've all got this faculty, and it needs to be sparked off with whatever trigger we find it useful to employ, whether it be drugs, reading horror, listening to Beethoven or going into the Peak District near here and climbing up a mountain. I attended a convention in the late 1980s in Huntsville, Alabama, not long after the Bible Belt had banned *The Wizard of Oz*. And they were talking about restricting children's imaginations. I found this attitude absolutely horrifying! What would anyone want to restrict the imagination for? The imagination is the most important mental faculty we've got. It frees us from the straight-jacket of subserviency and is the door through which we can explore the unknown.'

Does the unnameable thing lie through that door?

'It seems extremely unlikely that through the door there is nothing but a lifeless void. In Olaf Stapledon's *Star Maker* it is humanity which becomes the alien. We then have a glimpse of what it really means. I think there will always exist the unnameable. It is a perspective on both the human consciousness and the cosmic consciousness. Inevitably they must coexist.'

Ramsey Campbell's novel, *Midnight Sun*, is an attempt to strip away the metaphors and point a finger at this unknowable, unnameable thing.

'What the main character perceives is a Great Cold – a sentient thing of ice. Yet this is only the merest hint of what is to come. There is a myth that the midnight sun keeps something locked away and dormant in the far north where it shines. Perhaps this is the alien we are all searching for.'

If only we could see its face.

3
Alien Dreamer

That night the creature came out earlier than on any previous occasion, and Willy knew it was becoming very hungry. He watched and waited all night, but nobody came, and in the darkness of pre-dawn the thing reappeared on the river's edge on the way back to its lair. Somehow Willy could feel its anger and disappointment and hunger, as if the creature was a part of himself. He leaned out of the window, straining his eyes, wishing he could think of a way to help his friend. Suddenly he froze.

The shape had paused on the edge of the water and, although he could discern only a black patch in the darkness, he knew it had seen him. In some way, alien to humans, it had become very much aware of Willy and it dawned on him that the mysterious entity was not his friend at all.

It began to inch its way up Ridgeway Street.

(*An UnComic Book Story*)

Science-fiction writer Bob Shaw is a big man with a soft Belfast accent which carries a dry, subtle sense of humour. But he is serious about one thing – his love for the genre which enraptured him when a boy and made bearable the provincialism of life in Northern Ireland. For seven years, he was convinced he was the only person there who read speculative fiction. Shaw was an alien in his own land.

'I never really liked Ireland. I think it has produced more songs about the beauty of Ireland than any other country has about itself, but usually you find that the writers of these songs have never been anywhere else. Possibly because I'm an SF fan, I tend to think of myself as a citizen of the whole earth, rather than of just one particular small plot.

'As a child I resented the fact that I would read a comic and see things advertised that were obtainable in England but not in Ireland. If you asked for something even slightly out of the ordinary, the shopkeeper would shake his head and say he would have to send across the water for it. I formed the opinion early on that I should cross the water too and cut out the middle man. I remember my first visit to England in 1950. I immediately felt more at home there than I ever had in Belfast.'

The Shaws 'packed up for good' in 1973 and left Ireland to settle down in England 'to escape the madness'. They had moved home once before to spend three years in Canada. Perhaps this explains the mid-Atlantic feel to Shaw's work.

Respected science fiction author, Bob Shaw believes it is very difficult indeed to create really authentic aliens. Although they may not resemble us physically, writers still endow them with human thought processes. To attempt otherwise would mean a sacrifice of plot, and a boring novel! (*Bob Shaw*)

After leaving the firm of Short Brothers in Belfast, where he worked as press officer, Shaw freelanced his work just about everywhere and became a full-time, professional science-fiction author at the end of 1975. During his career he has published over 70 short stories, 20 novels, and numerous articles. The atmospheric piece at the start of this chapter illustrates effectively a moment of recognition between alien and human before it is overcome with fear.

Apart from Britain and America, Shaw's work has appeared in South America, Mexico, France, Germany, Italy, Holland, Spain and the USSR. He has won two Hugos and lectured in 10 different countries.

He is best remembered for his 'slow glass' idea, which saw its genesis in a story called *Light of Other Days*, and his novels *A Wreath of Stars*, the classic *Orbitsville* and *Orbitsville Departure*, and his paranormal fiction *Dagger of the Mind* and *Fire Pattern*. His recently published inventive trilogy, about a medieval culture on binary worlds a few thousand miles apart, proves he is still up there among the best. Recently a third novel appeared to complete the Orbitsville saga.

Why is the human race so fascinated with the theme of contact with extraterrestrials?

'From Stone Age man we have looked up at a starlit sky with a sense of wonder. Now we know the stars are distant suns, many of them with their own planetary systems, some of which must harbour intelligent life. Anyone with a bit of imagination must have visualized these alien worlds and wondered what they looked like. But more intriguing than the worlds are their inhabitants. A world able to sustain life wouldn't be all that different from our own, but the aliens could be totally different.

'That's the main interest in science fiction – speculation concerning our interstellar neighbours. It must be the most popular theme in the genre. Since the advent of the scientific culture in the last century, when scientific advancements began to make an impact on society, people began to realize all sorts of things were achievable – even communication with extraterrestrials. H. G. Wells triggered most of it off with *The War of the Worlds*. Mars is our nearest planetary neighbour, and in those days it was a fascinating proposition that across just 35 million miles (56 million kilometres) of space we might find our 'brothers', or our enemies. The opening of that book describes the Martians as 'cruel' and 'unsympathetic'. This has sent shivers down the spines of generations of readers. No wonder the concept of alien contact is the most popular idea in science fiction!'

First-contact stories still seem to have as much appeal today as they did 60 years ago in the pulp magazines.

'The cinema is to blame for this, but it relies on action and special effects. Real SF doesn't transfer to the cinema very well because it is mostly cerebral. It's stuff you're thinking about, the transmission of ideas – subtle ideas, difficult ideas, which don't really work in terms of moving pictures. But no matter how bad an SF film is, no matter how awful, there usually features at some point a ship landing on an alien planet. That first glimpse of another world always grips me. It's a glimpse into the unknown.

'One of my favourite films is still *Forbidden Planet*. It's that moment where they land and the doors open. You see the alien world, and from the distance something is approaching, sending up a cloud of dust. The suspense builds up. Who or what is arriving?'

There's another memorable moment when the viewer first sees the alien. Usually after that, a film goes downhill. A case in point is *Aliens*. In some ways this is a more polished version of *Alien*, except in one vital aspect: the second film lacked the impact of the first, because by then we knew what the alien looked like. It held no surprises. And strangely, both films ended exactly the same way. But why do you think the film *Close Encounters of the Third Kind* was the huge success it was?

'Basically because it went back over the whole von Dänekin – Bermuda Triangle, someone's up there looking after us – theme. I hate to sound like an élitist, but I think it's a great failure of our educational system that if you stopped the man in the street and asked him a few simple questions, about things I consider to be the cornerstones of knowledge, such as what is a star and what is a planet, he usually does not know.

'Perhaps we're all hoping to find the ultimate aliens somewhere, who are going to tell us everything we wanted to know about ourselves, do something for us we've always wanted done.'

Do you think this is why religious organizations, like the Aetherius Society, came into existence? They believe their leaders are in psychic contact with spiritual beings who exist on the other planets of the solar system. We tend to think of aliens as being superior either in physical strength, technology or spiritual attainment.

'I think it is a case of people casting around. Most of us are worried about the way the world has gone during the last hundred years. The old source of comfort was orthodox religion. But we've become increasingly more materialistic, and it's hard to believe that all the events in the Bible are true. As an SF writer I wouldn't try to get away with the scenario of the Christian religion!

'The new rising star was science. That was going to solve all our problems. Now people are disillusioned with that too. For too many people it has done the opposite it set out to do. So the logical next step is a combination of the best of both. A futuristic scientific element in the guise of the arrival of beings from another world, and a religious one, in the sense that we should worship them or fear them.'

In the context of fiction, aliens were originally portrayed as the aggressors. They came to enslave; they came to destroy and conquer. As far as the film world is concerned, this is still the case.

'I think it is probably a reflection of world history, and our own experience of the human race. Particularly when the world was opened up in the fifteenth and sixteenth centuries. The European explorers only went for one reason: to see what they could get out of it – to plant the flag, bring back gold and dispose of anyone who objected. I think down the generations this built up into a huge collective subconscious guilt, which has been exteriorized in science fiction. When scenarios were painted of alien ships landing on earth after travelling vast, interstellar distances, writers could only envisage that the reasons for these journeys were the same as those of our ancestors – to take over and subdue the natives, clean out all the goodies, destroy the 'primitive' culture.

'As the world's general political climate changed, colonization became untenable, and it was recognized as an evil desire. Then the literature began viewing aliens in a different light.'

Does that mean that the extraterrestrials in science fiction are really human beings in disguise?

'Yes, in spite of a writer's best efforts, the most determined efforts, usually aliens turn out to be humans dressed up as something else;

Bob Shaw's aliens are nearly always humanoid. This superb illustration for *The Wooden Spaceships*, by Chris Brown, depicts an aerial battle between the aliens of a binary planet system linked by a common atmosphere. (*Gollancz*)

physically different, but with mental processes which are the same. When I was a kid I read a story by an American called Stanley G. Weinbaum, who died before he got into his stride. It was one of a series of stories which were collected under the title *A Martian Odyssey*.

'In this particular tale, an earthman arrived on a planet and found some intelligent vegetables. He is having a deep philosophical conversation with one of them – as often happened in SF in those days – when an animal appeared which began to eat the thing. The earthman was horrified that this intelligence was going to be destroyed. "You're being eaten up!" he cried. "Yes, I know," the vegetable said, "but I don't mind. It's OK. It doesn't matter."

'It really hit me that here was a truly alien personality that didn't share with us a fear of death, and a drive for continued existence. There was something of the nursery tale in it – but it clicked. In the 50 years since then, I've rarely come across the same spark. I'm a professional SF writer, and have been for many years, but I have to admit defeat when it comes to the business of contact between aliens and human beings. They're from another world, supposedly totally outside our experience, yet for a story to work, and for me to invent an alien character, I cannot help but impose on it human characteristics.

'If, while walking through the jungle, I was confronted by a cobra, I would be frightened out of my life. I would run away with horror. There would never be an exchange of ideas between me and that cobra, even though we came from the same planet, the same evolutionary background. But you describe an extraterrestrial coming across a million light years from a totally different environment meeting with us, and for the purpose of the story it is necessary to exchange ideas.

'By definition, this being must be more alien than the cobra, yet he can speak English and handle the subjunctive like an Oxford don! I often feel I've fallen down on that point. Time after time in a story where I've known I'm going to be in that situation, I've sworn I won't take the easy way out and use telepathy. When I get there, I think, Oh God, what am I going to do?

'I had this idea of a "universal translator" – a device worn round the neck which could instantly analyse and translate any language."

In a desperate attempt to portray an authentic meeting between an extraterrestrial intelligence and human beings, Shaw wrote an experimental novel called *Palace of Eternity*. In doing so, he had to sacrifice a lot of traditional plot structure, and the critics did not take too kindly to it.

'There's no communication at all, because their minds are too far apart! In the story, there is a Foundation set up on Earth whose sole purpose is to find a way of communicating. Just to exchange one simple word would be seen as success. The project runs for centuries, and fails . . .'

This, perhaps, brings home the futility in our search. Shaw's idea parallels nicely the work of radio astronomers involved in listening out for radio messages from Out There. Listening for that briefest of message – just like Bob Shaw's Foundation, hoping for that one word.

'I thought it was my most realistic and honest attempt at contact between Us and Them. But I've done it once, and from a dramatic point of view it isn't too good. You can write a very thoughtful, deeply intellectual book about this non-communication problem, but that's it once it is done. You can't do it again.'

Is that why a lot of your aliens tend to be humanoid?

In his Land/Overland trilogy, which begins with *The Ragged Astronauts*, the aliens are indistinguishable from us.

'In those books, and I've done it quite often, I take care to explain the galactic background. In that scenario, at some time in the distant past, there was a huge explosion of a humanoid race which populated thousands of planets. When their culture collapsed, all that was left were isolated pockets of humanity. It is probably not all that unrealistic. Estimates of the age of the universe vary from between 10 to 20 billion years. That's plenty of time for some wandering race to spread their genes all over the place.'

Certainly there are passages in the Bible which could be interpreted in those terms. Have the aliens already been here? Are *we* the aliens?

'There are people, like Erich von Däniken, who have made a lot of money trying to convince others that is the case. But I'm afraid he has blotted his copybook too many times, although to his converts this doesn't seem to matter. If someone has this need to believe, once the idea has been planted and it becomes internalized, it makes no difference at a later date if another person comes along and says it is wrong. They've taken it on board and they're not going to let it go. It's the basis of all religion. There's an old saying in Northern Ireland, you can't reason someone out of something they haven't been reasoned into.

'There's a great cop-out in religion when you ask a tricky question: you must have faith. But that's what confidence tricksters say – trust me.'

Does that mean that you view actual alien contact as just fodder for your fiction, or is there a real possibility?

'I view the idea of physical contact with an alien race as a genuine possibility in the future. All the mathematics say there must be someone out there. Making contact is another matter, because of the vast distances.'

Granted it's out there somewhere, what are the chances that extraterrestrials will be humanoid?

'I'm not a biologist, but then perhaps that's a help. From what we know about the way intelligence develops, an alien life form is bound to have many similarities with us. Take the face for instance. I would expect an alien being to have its eyes above its mouth, so when it's eating it can keep a look-out for predators. If its eyes were below its mouth, it wouldn't be able to see during feeding, and its chances of survival would be low. It seems reasonable that ears should be either side of the head to obtain the maximum separation possible to pin-point the direction of a sound – another survival characteristic. There are a number of factors like those which are necessary if an organism is going to survive and progress towards greater evolutionary goals. Having fingers is not only essential for tool-making skills, but they are fundamental to the development of the cortex of the brain, for intelligence to be able to manifest itself in the first place. When we do eventually encounter aliens, they won't look like slugs or jellyfish, but just like ourselves.'

4
Alien Thinking

I suppose it's fairly prosaic if the dead come to visit you, but what if the people in your dreams are coming to say 'hello' from the more distant reaches of the Galaxy? ... While the more 'down market' of the UFOlogists carry on trying to persuade us that flying saucers are spaceships and that their occupants are either intergalactic tourists or evil beings from inside the earth (which is hollow, you see), the more serious investigators seem almost unanimous in their suggestion that the whole phenomenon is at root a psychological or parapsychological one.

(*Dreamers*)

John Grant (the pseudonym of Paul Barnett) sits astride the borderline between scientist and science-fiction author. His university grounding was in astronomy and physics, but he left all that behind when he went into the world of book editing. Here the time dilation effect is not so much a question of quantum mechanics but more the distinction between the normal working day and the seemingly endless editorial lunch-break.

He cannot be accused of being a stick-in-the-mud since branching out into writing himself. His novels have ranged from the alien disaster genre, *Earthdoom* (which he hopes to turn into a radio series), to two works that must qualify in anybody's list as the most imaginatively titled books in history: *The Truth About the Flaming Ghoulies* (the story of a super-powered rock group) and *Sex Secrets of Ancient Atlantis,* which masquerades as a pseudo-Erich von Dänekin report on certain mystical 'discoveries' that, in this instance, ought not to alter anyone's concept of the universe for longer than it takes to read the book itself.

At the other extreme, Grant is the author of several factual works that look at the limits of human and scientific theory, such as the highly illuminating *Directory of Possibilities*, which he penned with Colin Wilson, *A Directory of Discarded Ideas* and *Unsolved Mysteries of Science.*

The opening quote comes from what is widely viewed as his most scholarly work. In *Dreamers* he discusses the nature and purpose of the dream in human society, and in particular the dream of other realms and how it has influenced thinking in a very meaningful way. He notes how dreams have been at the heart of many great literary works (including early science-fiction tales like *Frankenstein* and *Dr Jekyll and Mr Hyde*) and also famous discoveries (such as the chemical nature of the Benzene

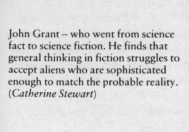

John Grant – who went from science fact to science fiction. He finds that general thinking in fiction struggles to accept aliens who are sophisticated enough to match the probable reality. (*Catherine Stewart*)

molecule). Nowadays the dream is equally revealing, he contends, in the world of alien contact.

We discussed the way in which science-fiction concepts of other intelligences have metamorphosed through the years.

'The depiction of aliens has certainly become much more thoughtful. Up to the 1950s and 1960s the average ET was either basically a human being with a funny body, a little thicko, or a monster intent on ravishing scantily clad women with large breasts. I've often wondered why a gelatinous beast, covered in suppurating boils and tentacles, should be remotely interested in a human female; surely the object of his lust would be a female gelatinous beast?'

As Grant points out, the change came about some 30 years ago when authors started to realize that aliens would not think or look like us.

'I particularly liked Ian Watson's portrayal in his novel *The Embedding*, where the ETs arrive on earth but are simply not interested in human civilization. They have been trading around the galaxy for aeons and have encountered many cultures, so all they are really interested in is finding out if trade with humanity is likely to be profitable. I regard Watson's idea as a very reasonable scenario.'

Nevertheless, Grant has his own peculiar interpretation of the question of first contact.

'I think that it's very likely that, when and if we establish contact with an ET culture, we may find that there's not a lot to talk about – indeed, that proper communication may prove to be impossible. Every civilization will have something in common, of course, but aside from basics, they will have evolved – surely – in very different environments and so will have quite different intellectual axioms.

'In fiction it is difficult to create a good plot around the idea that humanity and the aliens might be simply so different that very little communication can take place. Damon Knight tried it in a short story many years ago in which humanity and the ETs cannot stand each other's presence. It's a good story, but he still allows a great deal of intellectual common ground. I am not certain this is viable.

'In 1989 I wrote a short story in which the ETs were very, very different. In fact, they'd died out altogether. But they had left behind some metal boxes inside which was technology capable of converting visiting species to that race's way of thinking, which was utterly alien to human modes of thought.'

Although the story has received great praise from leading science-fiction writers, Grant has been unable to sell the tale.

'I think this is because SF in general is unwilling to accept that all the previous depictions of comprehensible aliens are based on less-than-stony-ground. So, all in all, I am not sure that SF's use of aliens has become sophisticated enough to portray the possible reality.

'A problem is that the only proven examples of intelligent life are those we know from our own planet. It is very hard to judge imaginative reconstructions against real science because real science has yet to cope with actual aliens of any description. The scientist lurking behind the science-fiction and speculative author has to ponder this.

'Recently I was chatting to Colin Wilson and he mentioned how cheesed off he was that reviewers of his Spider World novels were picking on the fact that giant spiders were a scientific impossibility. My feeling is that this displays narrow-mindedness on the part of the critics. When the science of aerodynamics seemed to prove that bumblebees could not fly, it was science rather than the bees that had to be modified – and a good thing too!'

However, he insists that scientific incompatibility is betrayed most in science fiction when it comes to alien thinking processes.

'They present the ETs as puppets who may look unlike us but think in the same way that we do.'

This, if it is not true – as Grant believes – represents one of the most fundamental obstacles to contact by any scientific method, because if we try to communicate with systems we regard as logical, there is a less than certain probability that 'they' will demonstrate any interest, awareness or understanding of them. Hence, contact via radio telescopes could be one massive waste of time and money.

'SETI is ridiculously underfunded. One reason why we haven't yet discovered evidence of alien civilization may simply be that brick-brained politicians refuse to accept that the search is of any importance whatsoever. It's like trying to find out about bacteria but refusing to use microscopes. The plaques we put on our probes that crawl out of the solar system are a waste of time. The chances of an ET civilization discovering one before the predicted death of the universe are very slender indeed; although I have a nightmare of an ET spacecraft at full speed hitting one.'

At this juncture there was pause for the tantalizing thought of a science fiction story commencing as weird alien visitors arrive, centuries from now, desperate to find some mythical creature called NASA in order to slap down a major intergalactic lawsuit and bill them for repairs and a mega-insurance claim!

John Grant did agree that the 'current methods of SETI are very anthropocentric.' We listen only to frequencies we think important and these are based on our body chemistry. 'What if ETs are ammonia-based?' Grant wonders. Then, presumably there would be aliens transmitting messages on frequencies we would never dream of examining because in our self-deluded arrogance we are convinced that anyone out there will be sufficiently like us to behave like us.

Despite recent funding of over a $100 million to the SETI programme, staggered over the next decade, Grant still believes it is not enough.

'Until the politicians are prepared to put more money into SETI, the search has only the most miniscule chance of success. It's as simple as that. The techniques in current use are governed less by scientific considerations than by financial ones. I get very angry about this, because I think that the unequivocal discovery of alien intelligence could trigger the most important scientific and sociological advance the human race has ever known – bar none.'

The strength of feeling that John Grant displays on this issue was very evident. He constantly referred to the need for more money to be invested but did suggest that an accidental breakthrough could occur.

'Some day a radio astronomer is going to come across an anomalous trace and discover that, lo and behold! here's the alien equivalent of a TV soap opera like *Neighbours*. The trouble is, that might not happen for another few billion years – if ever. An alternative could be that the aliens might think it worth the effort to come here in person. At the moment it is certainly not worth their while beaming messages at us, because we are just not listening.'

Once more we found ourselves straying from the purely scientific evidence of SETI into the realms of UFO abductees and 'channelling'. What does John Grant think of this reputed proof for alien contact?

'I have got to confess that I think it's all rubbish. Think about the "little voices" heard by the mass-murderer Peter Sutcliffe, which told him to go out and kill young women all over the north of England. These were a product of his mind, but as far as he was concerned they were a reality. I do not believe the "channellers" are trying to deceive us in any way. I think they are reporting things that have genuinely happened to them. I also do not think they are nuts. Everyone's brain occasionally gives a false signal – you have only to look at the experience of dreaming to realize that.

'Reality is a matter of perception: if our brains tell us something, then we obviously believe it. In *Earthdoom* the aliens offered an uncompromising message: "You earthling scum are the dregs of the universe. We come to annihilate you painfully and rape your planet." Of course, that was a joke, but I think that if ETs are going to contact us, they will make sure their message is just as unambiguous. I cannot really believe that they will pussyfoot around sending us confusing telepathic messages.'

Unless, of course, to an alien way of thinking what we imagine to be confusing telepathic messages are really blindingly logical and deeply meaningful contacts!

Not to be foxed by this rejoinder, Grant pointed out:

'Yes, but even if their own patterns of thought are very different from ours, in order to send a telepathic message in the first place they would have had to discover the basics of human psychology. So they would surely tell us something that was not open to misinterpretation, like where they came from.'

Grant has strong views on the influence that UFO contactees and UFO biographers of the last 50 years have had on society.

'I think a lot of people either misinterpreted what their brains told them or were out for a quick buck. I could name a more recent author whom I would put in the same category, but I won't. My belief is that these badly written, pulp SF fantasies did untold harm to the sober study of SETI and, indeed, of UFOs – whatever they are!'

He does not feel that this torrent of books is deeply influential on society.

'Actually, I feel that society created the books. People have always seen lights in the sky and produced explanations for them that relate to state-of-the-art technology. We have seen witches on broomsticks, phantom airships, now we have alien spacecraft. I believe in none of these, although I do think it is very likely that there are real lights in the sky that have yet to be explained satisfactorily.'

Why do people believe things that now seem scientifically unacceptable?

'I don't know, except that people believe what they *want* to believe. I have even had a letter or two from convinced students of Atlantis suggesting that my own [wildly exaggerated and improbable] novel *Sex Secrets of Ancient Atlantis* is basically OK but that I had made the odd mistake. The mind doesn't just boggle at that . . .'

If all this is true, then presumably the state-of-the-art will dictate that the alien spacecraft motif will eventually be replaced as an explanation for these lights. We wondered what the new solution might be so that we can make an early start and create a whole new branch of lucrative literature in the spirit of the great British entrepeneur.

Grant thought for a moment.

'In a decade it might be self-exploding Yuppies!'

5
William Loosley – Documentary Proof?

As we came to the clearing, I saw what I should have seen on earlier visits: but before I had been searching for scorched ground, and scrutinizing the grass a little at a time. Now my vision swept over the entire clearing, and I perceived the sign whose very size and breadth had allowed it to escape me: the grass and weeds lay bent and slightly flattened, showing where they had been pressed down over a wide expanse; I judged it circular; pressed down by some weight the size of a house.

So writes one William Robert Loosley, purporting to describe his encounter with a 'metal carriage' from another planet that came along with a 'fallen star' at Plummers Hill, near High Wycombe in Buckinghamshire. This close encounter was remarkable in just one way: Loosley died in 1893 and his out-of-this-world experience reputedly occurred very early on 4 October 1871 – thus making it the oldest alleged UFO landing in contemporary history, as well as the oldest one to which an entire book has been devoted.

The book is a slim, 96-page volume with the archaic title of *An Account of a Meeting with Denizens of Another World 1871* – based, apparently, on the Loosley manuscript around which former Oxford graduate physicist and science-fiction author David Langford added his 'editing' and 'commentary'.

Langford appeared fascinated with the scientific information imparted to the builder and cabinet-maker by this alien probe. It was twentieth-century physics far beyond the capabilities of even the best Victorian professors and allowed Langford to conclude the book with:

The manuscript of William Robert Loosley is clearly not a fabrication on his part ... The possibility of a later hoax is the next obvious one, and naturally this is less easy to refute: I can only declare that the manuscript has so far withstood every test of authenticity to which it has been subjected.

Do we have the first glimmer of hope that the dream of life on other planets has finally come to fruition? Many people seem to believe that we have. While the book has not exactly set the academic world aflame, its reception among the more esoteric minded has been welcoming.

It earned itself a favourable write-up in one of the best-selling UFO titles of recent years, which is still entrancing readers around the world: *The World's Greatest UFO Mysteries*, compiled by two journalists, Nigel

Writer Whitley Strieber grabbed the headlines with his book *Communion*, published in 1987, in which he sincerely claimed abduction by 'non-human entities'. Strieber received a severe bruising from media sceptics, but his book, and its sequel, *Transformation*, became best sellers. (*Peter A. Hough*)

Blundell and Roger Boar, written from newspaper clippings and not interviews. To be fair, though, the authors never pretend otherwise and evidently fulfil a need within the market. However, as can be attested by those of us who have been unfortunate enough to be interviewed for the sort of media tales this book reprints, the ultimate facts about a case of alien contact rarely find their best outlet in this manner.

Nevertheless, the promotion of the Loosley story in Blundell and Boar's book was influential. It subsequently became incorporated into the newest offering from Whitley Strieber, who has rapidly become established as the best-known international writer in the UFO-related field. His recent work, *Majestic*, is presented as fiction. It blends factual incidents from UFO lore with his panache for horror fiction in a clever reconstruction of alien contact. He incorporates the Loosley manuscript into his text, although having allegedly had no contact with Langford regarding it. When Langford read *Majestic*, he questioned Strieber's publishers (especially as he wanted a credit to his scholarly book to be included in future editions). Langford learnt that Strieber's source was reputedly *The World's Greatest UFO Mysteries*, but that he had been unable to backtrack further than this in his research.

What is most fascinating is how the chain of writing, from Loosley to Langford to Blundell and Boar and finally to Strieber, introduced a few distortions into the story. This happens often in such situations. It is almost inevitable when differing styles juxtapose, squeezing the evidence somewhere in the middle. No fault is implied on anybody's part because of this natural process, but it is interesting to see the result in this case and ponder its implications for other documentary evidence of alien contact that we find regularly filtering through into the media.

One distortion concerns the manuscript's discovery. The Langford book says it had been found 'a few years ago'. These words were written in 1978, but Langford says there are clues in his text which show that this discovery had to be after 1975. Strieber incorporates its discovery into a story which begins in 1947!

Then again, remember that final quote from Langford about the manuscript having withstood 'every test . . . to which it has been subjected'? Strieber reports that it has been verified by scholars in Britain; which is a minor, if subtle change of emphasis which was probably innocently stimulated by the tale's transition through the Blundell and Boar book.

All of this may seem of little consequence. After all, Strieber does say his book is a novel and so he need not necessarily believe the Loosley tale. Nevertheless, as Langford pointed out, only one of the two published descriptions of the document allows for the possibility that the number of tests 'to which it has been subjected' was zero!

As we were compiling this book, another development occurred. We chose the opening quote from the Loosley manuscript because it ties in surprisingly well with a phenomenon that has become a global passion during the past few summers. This is the so-called mystery of why swirled, flattened circles are turning up in abundance in crop fields around the world – but mostly across southern England.

While the sensible, scientific evidence suggests that a down-to-earth but quite interesting atmospheric phenomenon is to blame (see Dr Terence Meaden's *The Circle Effect and its Mysteries* and Randles and Fuller's *Crop Circles: The Mystery Solved*) there have been those who prefer a more dramatic solution.

The best-selling title which has triggered much of the interest comes from former *Flying Saucer Review* consultants Colin Andrews and Pat Delgado: *Circular Evidence*. Its authors reach no conclusions but have mooted many strange possibilities in their other writings, including symbolic alien messages that may be warning the earth.

By a fascinating twist, the modern crop-circle mythology began in Britain in 1980 – shortly after the Langford book appeared. The landing marks that Loosley claims to have discovered are closely akin to the circles that now plague British fields. What is more, a key site for these is by a hill near Aylesbury in Buckinghamshire, where several spooky happenings have been reported, which is also very near the spot where Loosley is meant to have had man's first contact with an ET . . .

On 30 January 1990 in the *Western Evening Herald*, a major newspaper in Devon and Cornwall, Mrs Marilyn Preston Evans – who has herself had some fascinating UFO experiences – wrote about this

A typical crop circle near Mansfield, Nottinghamshire, in July 1989. A message from the aliens, as some contend? No – in this case as in many others – a hoax. Other, genuine circles may have an atmospheric explanation. (*Jenny Randles*)

curious link. She wanted to 'point out the fact that [in 1871] . . . a spacecraft landed at High Wycombe in Buckinghamshire . . . causing a circle to appear'.

Aside from the problem that this statement of 'fact' takes several leaps of deduction, assuming for example that what Loosley saw really was a spacecraft and that the marks he reported finding were in any sense related to what he saw, if indeed he saw anything, Mrs Evans goes on to tie in the modern crop circles and a curious light filmed by an infra-red camera near one of them. She asks whether this was a 'sky-ship' like William Loosley encountered.

This speculation generated a response from a local sceptic, Lawrence Harris who, on 21 February 1990, argued powerfully for a logical solution to the crop circles and merely noted that 'A "report" by an 1870's builder would hardly contain meaningful scientific observations and simply cannot be used as "proof" of alien visitations.'

He was right to be cautious, but Mrs Evans had a valid point of view and had noted an interesting connection. Also Mr Harris had apparently not read the Langford book because it does include 'meaningful scientific observations'. Without them the story would be instantly forgettable.

Armed with an interesting article about his book written by Langford himself and published by *New Scientist* magazine (26 May 1988), but which few fans of William Loosley seem to have perused, we decided to discover the truth of this matter. There was a fairly simple way to achieve that — find David Langford. Given his prominence in the science-fiction genre and his many factual and fictional books, it did not prove that hard to track him down. He happily agreed to reveal all.

But first we decided to explore his views in general. Did he believe that the SETI programme to find extraterrestrial life was being correctly applied?

'I am subject to the gloomy fear that there might be nothing to hear out there after all: the usual worries about the possible rarity of life and in particular of radio-using technological life, compounded by the uncertainties of overlap.'

What did he mean by overlap?

'If a civilization is 40 light-years away and they are just building high-powered radios, we have 40 years to wait. If they used up their resources and gave up civilization one fleeting century ago, then we are already 60 years too late. But I would love these qualms to be proved dead wrong, and how can it be done except by spending money? Not that it is much use putting forward this argument in Britain, where miserable research funding currently means a loss of some 1,000 scientists a year to other countries.'

Of course, the urge to spend is confounded by the lack of probative results.

'There isn't any evidence as such — just some hopeful statistical indications. We are still embarrassingly short of data on how precisely life got going on earth, which seems to be of huge importance before any meaningful estimate can be made about the general case.'

Langford, a man always fond of a joke, likened the chances of success in SETI as rather less than his of winning £1 million on the British football pools. But he added:

'Think of all the millions of people who do the pools and other national lotteries. Maybe that's the way to present it: the Galactic Lottery, First Prize — is a complete new package of cultures and technologies whose value is literally priceless!'

This implied importance of the search, but feeble likelihood of success through conventional means, seems to indicate that science ought to show more interest in the less conventional evidence, such as the Loosley manuscript. What were Langford's overall feelings about the accumulated evidence of alien contact?

'I have to admit the doubtless very uncharitable view that if a phenomenon seems scientifically inexplicable, and if evidence for its reality is not available or testable, and if someone somewhere is nevertheless making big money out of it, then in terms of objective reality it's probably a load of rubbish.'

Here, he was referring primarily to the phenomenon of channelling alien messages via psychic contactees.

'When an alleged channeller or alien contactee relays a one-page proof of the four-colour theorem, or the answer to any other major and unsolved mathematical problem, I am prepared to be very impressed. I strongly suspect that any other civilization of comparable level which reasons at all will, one, use mathematics and, two, be ahead of us in some areas and lag behind in others – giving opportunities for two-way knowledge trading.'

As for the UFO phenomenon itself, he has equally specific views:

'My feeling is that it would be wrong to deny reality to this complex of reports and legends, but also that most of the true events were happening inside people's heads.'

Some contactee tales were probably get-rich-quick schemes, but others 'saw lights in the sky and, even in the most reductionist, sceptical view of reality, it still seems an overwhelmingly strong a-priori hypothesis that many lights will not be immediately explicable to many observers – allowing time for wonder and imagination to work. Some of the lights would not have been explicable even to a crew of trained scientists with full recording equipment. We don't know everything!'

However, while many witnesses are convinced, Langford seems reluctant to dispense with the opinion that in some cases the alleged alien contact experiences were really science fiction rather than science fact. He wonders if 'some people resorted to [UFO] books and found a waiting set of assumptions about flying saucers. Hey, I'm someone special, not everyone sees these things and this became – before long – in some cases actual memories that were coloured by what was in the book. Psychological support would be provided by the usual mix of perversity and self-interest known as human nature. One knows something that those overpaid government scientists do not . . . etc.'

Of course, these traditions survive into the 1990s with claims of terrifying abductions by weird creatures. What of these?

'It all seems very dubious to me. People who 10 years ago would keep ringing the police complaining about mind rays from the Chinese communists coming through the bedroom wall can now read the forceful template created in the wake of recent publicized cases and put it all down to a previously unsuspected alien abduction. Why should people fantasize such unpleasant experiences as some of those reported? Well, I am sure that being raped by the alien equilavent of a telephoto lens would be a jolly nasty experience; but if you don't actually have to undergo it but only "remember" undergoing it, suddenly it can be seen in a new light as a partly masochistic – or not so masochistic – fantasy. Overwhelmingly, it is a terrific device for getting attention and sympathy.'

This might all seem very curious to anyone who has read Langford's account of the Loosley story. Here he has presented us with some of the best available evidence for the reality of alien contact via a UFO; now he seemed to be disputing the objective reality of almost all of it.

'Up until 1988 I would tell anyone who enquired that I myself remained very sceptical about the manuscript. Since then I have invariably referred them to the "definitive investigation" in *New Scientist*.'

In fact, this short piece by Langford was originally intended for the *Skeptical Inquirer*, the house journal of a group of scientists, writers, magicians and the like who are ultra-critical of any claims regarding the paranormal.

The article was an outright confession by Langford that he had hoaxed the entire Loosley story; something which had probably not come as a major surprise to any researcher who read the book with much care was familiar with the author's notoriously wry sense of humour.

His brief from the publisher was to write a spoof book about a nineteenth-century UFO encounter; to examine the evidence as a physicist would; to lambast modern ufology for its lack of scientific rigour; and to make the Victorian UFO sufficiently over the top that no close reader could believe it. Langford succeeded admirably, although some readers clearly did believe it.

Such a wheeze has its good and bad points. Jenny Randles was forced to explain the truth to one elderly lady who wrote about the story in some detail. After all, it did seem to challenge the opinions in Jenny's own books which suggest that UFOs may not be spacecraft. When faced with the truth, the elderly lady reacted with understandable angst, saying that as an old-age pensioner she felt cheated out of her money (although the book is a darned good read). She asked us to suggest to Mr Langford a graphic proposition of what he could do with his hard-backed book!

However, we would report that David Langford reacted very honourably and, with our help, the lady was recompensed, although this is not guaranteed as a general rule. Nowhere on the book's cover is the reader misled by any statement that it is a true story. Loosley did exist, as a distant relative of Langford's wife (as clues in the text actually indicate). The failure to expose the hoax was largely the result of a general ineptitude, since despite various media reports on the case, none of the journalists bothered to request a sight of the century-old manuscript. This mysteriously fails to appear among the assorted faded illustrations of Loosley's shop, his family, a gravestone, etc., that are included in the book.

However, to the credit of serious ufology, it must be said that we could not find a single source here that took the novel seriously.

It is fascinating to compare such caution alongside the credulous way many non-ufologists accepted it. A detailed review by Colin Bord in issue 31 of *Fortean Times* magazine (Spring 1980) was typical and predictive. He pointed out numerous problems, including the missing photographic proof and concluded:

It may be perhaps that I am taking too solemn an attitude to what is only intended to be an academic joke at the expense of ufologists. But too often in the fields of ufology and Forteana it has been shown that this year's tall story became the future's factual report, passed on from one hack writer to another and accepted as the truth by thousands of gullible readers.

Bord demanded that the publisher or author provide the proof requested or own up to the hoax; neither was done for eight years.

This little jest by David Langford might be perceived as an interesting sociological experiment, making us question the validity of all evidence of alien contact. To what extent is this necessarily tempered by our all-too-human desire to want to find proof? If we believe in extraterrestrials enough, then, like the proverbial fairy whose very life depends upon the willingness of a child to believe in her, might we not actually stimulate their existence in a metaphorical rather than tangible sense?

David Langford reacts to all of this in a philosophical manner. He said with a penetrating smile:

'My finest hour came when I was attacked for the excessive caution and scepticism of my own commentary on Loosley's narrative.'

It would seem that the physicist within him yearns for scientific evidence of SETI that as yet has singularly failed to arrive. So, for a time at least, the science-fiction writer was able to invent his own proof and share it with a world that desperately needed to believe that it was true.

6
Aliens on Film

In this first section we wanted to survey the opinions and attitudes of those who have the facility to 'dream' about the existence of alien intelligences for a living. One thing we noted was the curious antipathy between those who study the supposedly factual evidence for alien contact (such as ufologists) and the community of science-fiction writers and fans. One might expect there to be much common ground, but it is an all too common grievance that the media lump the two together. In fact, many science-fiction writers are passionately opposed to the view that UFOs are extraterrestrial spacecraft and look down with considerable disdain on anyone venturing 'evidence' that they are.

Others have little to offer on the debate, regarding what they do as an imaginative exercise somewhat divorced from reality. Fantasy author Mary Gentle was a case in point. She was willing to state for the record her views on the existence of extraterrestrial intelligence, which were simply put: 'Quite honestly, I don't have any opinions on the matter!' – itself quite notable.

Of course, serious science-fiction literature is often very different from the Hollywood movie interpretation of the genre; rare indeed are the cinema releases which treat the topic of alien contact with great intelligence. The moguls who fund the films seem to believe that, in order to attract the science-fiction audience, you need to take Robin Hood and Maid Marian, dress them up in spacesuits, replace bows and arrows with laser guns and pit them endlessly against the baddies in an interplanetary shoot-out.

Usually the result is tiresome tosh, but occasionally, as in *Star Wars*, it can be entertaining – so long as you do not let any stray thoughts of scientific fact and credibility get in the way.

That you can make excellent, even gripping, science fiction movies that do reflect accurately the creativity of the written word is shown by the superb, atmospheric rendition of the films *Alien* (1979) and its more violent sequel *Aliens* (1986). Here the mentality of the alien predator is just as distanced from ourselves as its horrendous physical form. The alien landscape is also wonderfully recreated on screen.

However, in general, truly representative films of science-fiction literature are less popular with a wider audience. For example, the very

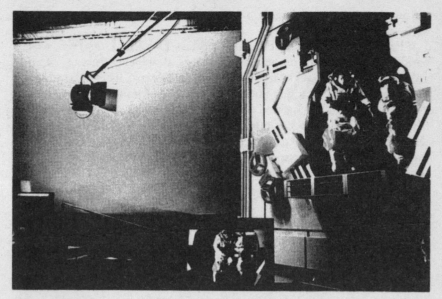

Hollywood's version of the universe. A space walk from the movie *2001: A Space Odyssey* re-created by Universal Studios, illustrating how special effects can provide a dream for millions to share. (*Jenny Randles, Universal Studios*)

clever ideas portrayed by the film *Quatermass and the Pit*, where an alien spacecraft is found buried in the London underground railway system, wreaking psychic havoc by unleashing the inherited traits of its alien originators, is little known. But similar, yet imaginatively bankrupt films which offer battles between warring spaceships do a roaring box-office trade.

The same cannot be said of a more recent offering, *Enemy Mine* (1985). This was released almost without publicity, but it makes a more credible attempt than almost any other film to portray the meeting of an alien intelligence with a human being, stranded together upon a hostile and genuinely unearthly abode. The special effects and alienness of the extraterrestrial and the planetary environment on which they are trapped are virtually unsurpassed, yet the movie gained little attention outside a cult science-fiction following.

Few people would doubt that one film above all others combined intelligent science fiction with cinema entertainment – *2001: A Space Odyssey*. That was the result of a joint effort between producer Stanley Kubrick and leading science-fiction author Authur C. Clarke, whose short story *The Sentinel* was also the stimulus for the script. Clarke went on to write a couple of sequel novels (one of which, *2010*, was also filmed many years later). He became an international celebrity because of the huge popularity of the original film when it was released at the height of the race to get to the moon.

What *2001* did, just like the more modest TV series, *Star Trek*, was to parallel our very real, dramatic technological achievements in stepping out in our own 'back yard' – the solar system. The film added a dimension which so far has eluded our space explorations: unambiguous discovery of alien intelligence. Such is the immensity of the universe and the apparent lifelessness of our neighbouring planets, the excitement of those first forays into space was badly tempered by the inability to grasp the real prize: contact with alien life. Critics say that this can only be a possibility in the dim, distant future, when our technology has reached the point of interstellar travel. In the meantime, films like *2001* remain an absolutely vital substitute. That film, in particular, probably boosted the imaginations of those who had the power to keep the space race alive and kicking.

In fact, as a story, *2001* leaves a lot to be desired. Many people have seen it a dozen times and still do not understand what the last half hour of the film means! But it was not the script that mattered: *2001* was a success because it lived the dream in the hearts and souls of millions. It took pure human emotion and dressed it up in dazzling special effects.

Despite being seriously ill in his retreat on Sri Lanka, Arthur C. Clarke was kind enough to offer us some comments. He, more than any other writer, may be fairly described as the catalyst for the dream that we discuss.

Clarke is quite candid in saying that there is no necessary connection between UFOs and the question of life in space. They may be entirely separate issues.

'Yes – I believe there is life in outer space. Yes – I think it is possible it could visit us. But I do not think that it has in historic times.'

UFOs, on the other hand, are, he believes, the product of three groups of witnesses: 'Liars, madmen and honest people fooled by unusual phenomena'. He seems fed up with the whole UFO field and says he has frequently spoken what he would hope to be his last words on the matter. Indeed, in his somewhat pompously titled book *Arthur C. Clarke's Chronicles of the Strange and Mysterious*, he claims that ufology sends him into 'uncontrollable fits of yawning'.

Clarke says that UFOs may sometimes be unexplained natural phenomena (for instance, he seems to accept Dr Terence Meaden's theory that those so-called 'mysterious' crop circles are not caused by 'alien beams' but are the product of atmospheric vortices). No UFOs represent alien spaceships, he states. In fact, he repels all comment on this opinion by quoting one of his favourite book titles, *Shut Up, He*

Explained! In September 1990, Arthur C. Clarke wrote to us on the subject of crop circles and this rational solution, noting 'It would be indeed ironic – and indeed poetic justice – if the crop circles, far from proving the E.T. hypothesis, drive the last nail into its coffin.

Many people believe that the best novel on the consequences of alien intelligence contacting humans is the ground-breaking *Fade Out* by Patrick Tilley. This is heartily recommended to anyone who is genuinely interested in how human or alien thinking processes might interact. Yet nobody has made a film of this wonderful story, although Tilley tells us that it very nearly happened! Plans were well advanced, when someone called Steven Spielberg decided to follow his highly successful tale of a shark attack (*Jaws*) with a fictional rendition of a subject that had fascinated him – the UFO phenomenon. Because Columbia Pictures were investing vast sums into this movie, the *Fade Out* project emulated its title and did indeed fade out. However, they would have been very different films.

Spielberg's *Close Encounters of the Third Kind* is in its own right a good example of how fact and fiction can be blended in an enterprising manner. It remains one of the highest-earning films of all time and one of the few which indulged in a re-release after extensive editing (including the shooting of expensive new scenes). This was a year after the original cinema screening and it was only possible because of its financial success and because at the time (1978) Spielberg could do no wrong in the eyes of those who controlled the purse strings.

In truth, *Close Encounters* is rooted closely in the actual reality of the UFO phenomenon (if these two terms are not mutually incompatible!). Spielberg relied heavily on the input from ufologists and some of the encounter scenes were based on genuine sightings.

The main consultant was Dr J. Allen Hynek, an astronomer who for over 25 years was official scientific adviser on UFOs for the US Air Force. When he left to found the (still-flourishing) Center for UFO Studies, he wrote what is regarded as one of the seminal works in the field, *The UFO Experience*. It was on this book that *Close Encounters* was quite firmly based; even the peculiar title derived from Hynek's classification system for UFO reports: an encounter of the third kind being one involving entities.

The late and much missed Dr Hynek even made a brief guest appearance in the film as one of the scientists communicating with the aliens. In one scene (cut for release) the creatures even stroke his beard.

This film was fictionalized, of course. It was also confusing in parts because it attempted to display both the psychic and material dimensions that underlie reports.

The biggest problem that *Close Encounters* illustrates is that it takes for granted the extraterrestrial explanation for the UFO phenomenon. It makes the logical, but incorrect, deduction that UFOs and spaceships are necessarily synonymous terms.

It may surprise many of the millions who have seen the film that its key contributors were never persuaded that UFOs are alien space probes. The French scientist in the film (played by the late François Truffaut) was based on Dr Jacques Vallée – Hynek's close companion and still an active contributor to the UFO debate via books such as *Confrontations*. Vallée has always been candid in his appraisal of the subject and he confirmed to us recently that 'I will be most disappointed if the UFO phenomenon turns out to be nothing more than extraterrestrial contact.'

Jenny Randles also knew Hynek well and he grew less and less convinced about the extraterrestrial nature of UFOs. Always the scientist, he favoured a more esoteric interpretation, but he was really just baffled by the contradictory evidence.

Around 1979, after *Close Encounters* was a success, Jenny recalls a lengthy discussion with Hynek as he canvassed ideas to take back to Spielberg for a then planned (but as yet unmade) sequel. The title suggested was *Close Encounters of the Fourth Kind*, since it would describe what happens when a human claims to go inside a UFO, as the main character in the movie had done at the climax of the film.

The main consensus view that developed from the meeting was that the sequel movie should not follow the expected pattern. Instead of taking character Roy Neary (played by actor Richard Dreyfuss) to another world somewhere in outer space, he should be confronted with the paradoxes of ufology and perhaps find himself descending into inner-space reality instead. Of course, whether Spielberg would have paid heed to these discussions remains unknown, but it was an entertaining debate.

In 1978 Jenny travelled around Britain helping Columbia to promote the film at cinemas and she also participated in a data-collection exercise in conjunction with the *Daily Express*, which had the newspaper serialization rights in the script. This conclusively demonstrated the amazing impact the film had on society, but it did not provoke waves of new UFO sightings as sceptics vociferously predicted. Instead it encouraged witnesses to report old sightings which they had previously kept secret for fear of ridicule. Nevertheless, people's thoughts were turned towards the universe, in the same way as the landing on the moon and 2001 had done a decade before. The motto of the film, 'We are not alone', became part of our culture.

It would have been very interesting to compare the *Fade Out* movie with *Close Encounters*. It is ironic that Patrick Tilley started off a chain reaction for successful (but hardly original) movie scripts involving prehistoric monsters in duels with strapping space heroes in the midst of a lost world; whereas his own true masterpiece of alien fiction remains uncommitted to celluloid.

Just what is considered acceptable in a film based on the alien theme tends to be dictated by what the money-makers think will sell. For example, Jenny has been in two sets of discussions about the filming of her non-fiction books. Neither have so far come to fruition; although *Sky Crash*, about a reported UFO crash near a US Air Force base in East Anglia, did sell a movie option and remains 'on ice'.

The other book, for which talks never progressed to an actual deal, concerned her aptly titled *Alien Contact*. This reached the stage where actors and actresses were even approached with a script outline, but nobody involved in the project was ever very happy with the plan to alter the whole character of the story.

The book describes the stories of some children who had had various strange experiences of a UFO-contact nature and assesses them from a variety of viewpoints. The concept of the film was to relocate the events in America, jazz up the experiences in such a way that the special-effects team could have a field day, and impart an unambiguous message that the experiences were the product of an extraterrestrial landing. However, in the real story we never knew whether the experiences were of an extraterrestrial nature and even the witnesses were puzzled about their interpretation.

A true film about this extraordinary family and their experiences would have to have their blessing and retain the essential contradictory, baffling and intangible nature of the affair. Yet this would not translate well on to the silver screen, where in order to be commercial it would probably end up a bit like *ET Meets the Muppets!*

One film producer who was prepared to give us his views directly was Robert Emenegger from Los Angeles. He masterminded the film *UFOs: Past, Present and Future*.

There are persistent rumours within the UFO community about some unknown Hollywood producer being called in by the Government to make the definitive film to demonstrate proof of alien existence so that the public would be 'educated'. Some even claim (although with little credibility) that Spielberg's *ET*, with its nice cuddly images, was an innocently manipulated part of the plot. Emenegger would only say that visitations by real-life ETs was 'Not only possible, but very probable, judging by the evidence I have seen'.

He thinks that aliens landed here 'as early as the twelfth century and throughout history'. Moreover, he says, enigmatically, 'from my information it happened in May of 1971 . . . other reports put it also on 30 December 1980 in England, both at military bases.'

The December 1980 date coincides exactly with the reputed incident at US Air Force Bentwaters and Woodbridge in Suffolk – the subject of Jenny's book *Sky Crash*. From her involvement in this complex affair (updated in the 1991 book, *From Out of the Blue*, published by Inner Light), she can state beyond any question that something came down, but it is rather unlikely to have been a spaceship.

The other date offered by Emenegger is more puzzling. It seems to refer to no specific case in his film or book but might well coincide with that rumour of a landing on the tarmac at Holloman Air Force Base in the New Mexico desert.

Here, film was allegedly taken by US Air Force personnel that proved beyond any doubt that aliens have visited. A mysterious unnamed film producer was – again according to rumours – later shown this in secret when it was contemplated that the news should be released to the world.

In Emenegger's book there is a fascinating short chapter entitled 'On Being Contacted'. This begins by saying 'Let us look at an incident that might happen in the future – or perhaps could have happened already.' It then goes on to describe an undated hypothetical event at Holloman Air Force Base where a UFO lands, aliens step out and communicate with senior officers and the entire thing is filmed for posterity. Emenegger then put 'the Holloman scenario', as he termed it, to several leading social psychologists and asked how this proof should have been handled officially and how society might have reacted. We will look more closely at the results of this detailed research project in the last section of the book.

It is very interesting to contrast Robert Emenegger's firm opinions on UFO landings with his views of other evidence for alien contact. He considers the contactee material from the 1950s quite bluntly as 'science fiction': modern channelling claims are 'from the self – not other entities . . . or, perhaps, hoaxes for money'; and 1990s' abduction witnesses to be perhaps mistaking something 'within the mind of the experiencer'.

Given all of that, one might well wonder why he seems so reasonably certain that alien life has visited earth on these quite specific dates.

Perhaps one day he will make a film that offers fictional insight into what everyone agrees would be one of the most momentous events in the history of our planet, an event that would, quite literally, change everything.

The Belief

7
Have Spaceships Landed?

According to a particularly strong belief within today's society, we have no need to ponder the abstract questions of life in the universe or dream about the possible appearance and motivations of any aliens, because those aliens have been here, right on earth, since at least 1947.

The UFO debate has generated more hot air than a room full of politicians – even the most basic issues provoke controversy and inflated egos add fuel to the fire. Yet within all of this barbed discussion, some fundamental questions do arise that require our consideration.

Unlike some paranormal phenomena, ufology has a grounding in hard fact. There are millions of witnesses from almost every nation on the planet who claim to have seen something inexplicable in the sky. Many of them are indisputably sincere; often they are rational, lucid and intelligent; occasionally they are even skilled observers, such as police officers, airline pilots, military personnel and scientists – even Clyde Tombaugh, the astronomer who discovered the planet Pluto.

Normally we would take their word about what they claim to have seen. The problem here is that what they claim to have seen, a UFO, is considered 'impossible' according to the conventional world view.

Dr J. Allen Hynek, whom we met in the previous chapter, used to say that ufology comprised credible witnesses seeing incredible things – and indeed it does. However, the overwhelming majority of those incredible things turn out to be misperceptions of mundane phenomena, thanks to the fallible nature of the human senses. Few ufologists would dispute that, although they may quibble with our contention (based on experience) that this accounts for 95 per cent of all reported sightings.

Consequently, we know immediately that almost all of the stories purportedly relating to alien spacecraft, with which the cheaper end of the media market overflows, are in fact evidence of nothing more exotic than the limitations of being human within a complex environment, because there are so many ordinary phenomena that can be mistaken.

This instantly poses the fascinating question as to why any sensible ufologist should contend that the dwindling number of puzzling cases are that proof of alien life for which humanity has so long craved.

Dr Shirley McIver, who received a Ph.D. from York University, in England, for studying the motivations and beliefs of UFO investigators

Dr J. Allen Hynek (centre) worked for the American government as an astronomical consultant on the subject of UFOs. He realized he was being put under pressure to explain away sightings when no logical explanation would suffice. Eventually, he resigned and founded the Centre for UFO Studies, which still flourishes, several years after his death. (*Peter A. Hough*)

(rather than UFO witnesses) wondered if the answer lay in an overall disenchantment with scientific omnipotence, a feeling that there had to be more to life than the professors decreed. There is no doubting that ufology adds a thrill to life. There are also hints that UFOs are a status symbol to the average person, since they can elevate their respectability in a peculiar sort of way. To say you believe in UFOs or have seen one can make you the centre of attention. Subconsciously this could be more important to the subject than any ufologist might be willing to recognize.

However, it would be wrong to dismiss the UFO field as the province of egotistical dreamers with delusions of grandeur. There are many highly intelligent and well-qualified researchers, from astronomers, such as Hynek, to psychologists, engineers, physicists, meteorologists and, geologists who are drawn by the evidence, not by the dreams. This evidence does at least have the superficial appearance of being persuasive; albeit, not necessarily of extraterrestrial contacts.

Hynek, for example, was involved with the data first hand for longer than anybody else. He saw the early American government projects from the inside and faced the cover-ups and secrecy they were forced to utilize. This was basically because nobody could work out precisely what was going on. To admit that ignorance – with its inevitable implication that the US Air Force believed spacemen to be cavorting about the skies as the authorities sat and watched with impotence – would have been detrimental to any elected government. A not so discreet silence was the consequence.

Hynek once told us a story about the time he visited the Pentagon (in the mid-1950s) and made one of his periodic attempts as scientific adviser to get an upgrade in their research into UFOs. They looked at him sympathetically, told him frankly that they could not be seen to be too positive because UFOs were a sensitive issue, but said, hinting perhaps at moves taking place above both their heads: 'Allen, do you really think we would do nothing?' Yet soon afterwards the American government tightened up its statistical juggling tricks and began to use new manipulative tactics to explain everything away. The officer in charge of the US Air Force investigation team was told, shortly before he quit in disgust, to 'stop mentioning the unknowns' and only tell people about those UFOs they had managed to resolve. Disinformation has long been the name of the game.

Such paranoia, understandable as it is, camouflages the real issues. Like most ufologists, those government scientists evaluating the UFO problem go through changes about as often as the moon alters phase. One month they pen secret reports suggesting that there really might be spacecraft on the loose; then they re-evaluate the data and wonder whether the seemingly solid evidence is more a product of wishful thinking than factuality. Consequently, research gets temporarily downgraded. Years later new personnel come fresh to the evidence and try to work it out, only to be faced with all the baffling paradoxes and ambiguities that fill each puzzling case. If they then conclude that there is a chance the aliens have landed, official UFO study is back in business.

There is accumulating evidence from government archives, statements by influential figures and data squeezed out from nations where Freedom of Information Acts have been imposed (on often unwilling administrations) that this state of schizophrenia persists. Nobody seems to know for certain whether UFOs prove the reality of spaceships. You can read the evidence in such a way that sometimes it is all rather insubstantial, whereas at other times it looks to be very probable. Given an attitude of 'we don't really think these UFOs are spacecraft, but they might be; we can't fully prove they aren't and we don't exactly know what else these darned things might be', a very confused and hushed-up monitoring process is the all-too-predictable result.

Hynek went through this same metamorphosis. Even near his death in 1986 and after a lifetime of research, with fame and at least modest fortune behind him, he was asked to star in a TV commercial to promote beer. He liked the beer, but he was debating whether to do it because the script he was expected to act out presumed UFOs to be extraterrestrial. Above all else, Hynek was an honest man and he knew that the evidence may be suggestive but did not prove that to the satisfaction of a scientist.

In 1984, having moved to the desert climes of Phoenix, Arizona, in what was as close to retirement as this great man ever came, he gave one of his last major interviews on his UFO beliefs. He told Pamela Weintraub of *Omni* magazine: 'My biggest hope is that as the research progresses, I will be able to demonstrate that the ET hypothesis is untenable, that it just doesn't wash to a scientist.'

So what did Hynek propose in place of this popular view? Like most ufologists, he was perplexed. Wherever you looked for an answer, you ended up with a hundred new questions instead. On one of his visits to Britain, we took him out for a meal to the Wizard Inn near magic-steeped Alderley Edge in Cheshire. We went there as much to pander to his lovely sense of humour (the 'funnies' – comic strips – were always the first part of any newspaper he turned to) as for its evocative location for an evening out.

Hynek loved this because he always likened the UFO phenomenon to the Cheshire Cat from *Alice's Adventures in Wonderland*. What he meant by this was that, just like Alderley Edge's wizard or that magical cat, UFOs flew through our airspace quite unlike any conventional aircraft or rocket ship. These have a specific arrival and departure point, with a pre-determined route in between – not so a UFO. This blinks into our reality, cocks a snook at a couple of witnesses as it sails by, sometimes even leaves tangible signs of its presence (such as landing traces or photographs) *en route*, then it simply vanishes as if it had never been there.

For this reason Hynek tended to favour the view that UFOs might be a manifestation of a sort of dimensional interaction. They could be the key to the nature of reality rather than the secrets of the cosmos.

However, he was only dismissive of the idea of UFOs as spaceships in a literal nuts-and-bolts sense. He was mindful of the famous truism of science that any alien technology would be so superior it would be indistinguishable from magic. Hynek never believed that a UFO was the Alpha Centaurian equivalent of Apollo 11 piloted by little green Neil Armstrongs. Yet he also knew that any truly alien intelligence might possess a technology or communicative facility that was so incredible we could not even comprehend what it was, let alone begin to duplicate how it worked. In that sense, UFOs could be a 'holographic implant', which was the only way our minds could visualize, in comic-book imagery, some sort of long-range communicative probe from another world.

This idea, formulated late in his life, was developed by Jenny Randles in her examination of the close encounter of the fourth kind, *Abduction*. It was dismissed then by ufology, just as it was rejected as Hynek's trail-blazing vision. However, despite the insistence of those pundits of ufology that the evidence *must* either reflect real spaceships or some wild fantasy, Hynek was aware of the major problems both these hypotheses have to face. He used to liken the UFO to the dilemma that bamboozled physicists when they came to understand the nature of light. In some experiments it behaved like solid particles and in others it had clear proof of being electromagnetic radiation. Which was it? The question bothered science for years until they stopped asking it. It was only a problem because their knowledge was not yet sufficiently advanced to accept that light was indeed a radiation field that sometimes manifested in the same way as particles within our vision.

UFOs do exactly the same and Hynek was one of the few people who really understood this. If you are stuck with the question are they this or are they that, the reason is quite simple: it's the question that is wrong.

The other extreme of ufology can be divided into two distinct groups. There are the debunkers, such as aviation journalist Philip Klass and space writer James Oberg, who won a prize offered by *New Scientist* magazine for a UFO article. In their work they usually argue that the remaining 5 per cent of unexplained cases are only unexplained through lack of data and poor investigation. There is some limited support for this view.

Dr Jacques Vallée (*front centre*), scientist, science-fiction author, ufologist and model for the character of Lacombe in Spielburg's *Close Encounters of the Third Kind*, addresses the United Nations on UFOs in a debate brought about by Grenada in 1978.

Hynek, for instance, employed a young astronomer, Allan Hendry, to work full time for the Center for UFO Studies, using all the resources possible, in an effort to solve every UFO case reported to the group during 18 months. His results were published in the classic work *The UFO Handbook* and he very nearly succeeded.

But overall he was still left, despite the unparalleled access to equipment, money and contacts, with over 100 out of 1,300 cases that were unexplained. Many of these were so fascinating, and featured witnesses of such competence, that it was considered improbable that they would ever be shown to represent misperceptions for which there was just not sufficient evidence to identify the cause.

Faced with cases such as these, the last resort of a debunker is usually to cry 'hoax', often with doubtful credibility. This is where the debunking school of thought most displays its ineffectiveness. While making many good points, contributing much to our caution and being right in a large number of cases, it risks all on a gamble – and in our view loses – when it seeks to remove the best cases by a foul because it simply cannot bear to have anything left unresolved.

The other extreme view is very popular outside America, especially in Europe. This is sometimes called the psycho-social hypothesis, but it is more correctly a sort of rationalist or reductionist approach.

There are some fine books in this literature, notably from English scholar Hilary Evans and many more in the French language from an entourage of thinkers, such as Michel Monnerie.

It seeks to account for the remaining close encounters that cannot be put down to misperceptions, by arguing that they are psycho-dramas acted out in the minds of the witnesses and often supposedly triggered by tensions that are current within society, such as the widespread fear of nuclear holocaust that was especially powerful in western society during the 1950s and 1960s.

There are a few neo-revisionists, like John Spencer, who in his book *Perspectives* suggests that the answer is psychological, but this may be a screen for deep awareness of some undefined 'other lifeform'.

The key reason why ufology is turning to inner space in its quest for answers is that there are overwhelming grounds to believe that the close encounter does occur during an altered state of consciousness. Witnesses involved also tend to have gifted abilities in visual imagery. This has been established from case histories, psychological-profile testing of witnesses, crucial little clues (for instance, a remarkable tendency for abduction witnesses to have extraordinary early-life recall: having visual memory of being a few months old in the cot or pram) and a number of critical points emerging from the data itself.

For example, there is a lack of physical evidence of anything truly alien, such as a bit from within a UFO or a photograph of a landed spaceship that passes muster. There are also several known cases where witnesses have been observed by uninvolved third parties during the time when they believed themselves to be aboard the alien spacecraft. In no instance were they seen to be anywhere but where they had always been. These third parties report the witness as remaining there in a trance, or in a deep somnambulistic sleep or unconscious on the ground. Surely this leaves scarcely any room for doubt that the alien contact takes places while the witness is in an altered state and cannot be a real-world experience?

Ufology defines this state of consciousness by a series of reported symptoms one can spot in the data, such as a lack of proper flow of time, loss of sensory awareness and induced paralysis. The term 'Oz Factor' has been coined to cover it because (as with Dorothy in the movie *Wizard of Oz*) the person is claiming to be temporarily whisked away into another realm but then returned with no evidence that he or she went anywhere.

Research psychologists are becoming quite excited by these ideas. For instance, Dr Kenneth Ring, from the University of Connecticut, extended his marvellous research into another peculiar altered visionary state: the so-called 'near-death experience' as reported by patients who survive after nearly succumbing to surgery. He found sufficient parallels to propose an 'imaginal interpretation' of UFO close encounters.

We have had some interesting correspondence with Dr Ring, who was largely unaware that ufology outside America had embraced similar views to his own for some time. He stresses that 'imaginal' is not the same as 'imaginary' and he is talking about experiences which are not 'the stuff of fantasy' but, in a fundamental sense, a sort of reality. The inner-state world is one 'that has form, dimension and most important for us, persons'. It is also a world 'somewhere between what we still call mind and matter' (see his article in the American *MUFON Journal*, May 1989).

His conclusions, after a perceptive discussion of the problems, have led many serious ufologists to reject the extra-terrestrial hypothesis (or ETH): 'the search for the alien somewhere out in the galaxy must be abandoned anyway, for he is not there. You will find him instead in the multi-dimensional richness of human experience on this planet.'

Of course, Ring's theory stirred up great controversy in America. Ufology in that country embraces the ETH far more readily than anywhere else and this was heresy to many commentators. They waved it away as pseudo-mysticisim in the manner in which they dismiss much

European reductionist thought. To them, Ring was seen as ignoring the physical dimensions which are perceived to exist and so was going against the basic tenet of the ETH. This argues that if people all over the world are reporting essentially similar encounters with aliens and UFOs, then the differences are largely irrelevant because the overall emphasis is in favour of a true alien presence.

It is important to look at some of the reasoning behind this argument. Perhaps the best person to consider is Jerome Clark, who took over from Hynek as editor of his society's journal, the *International UFO Reporter*, now widely considered the leading publication in ufology.

Clark was one of the first American researchers to turn away from the ETH toward what at the time was called 'New Ufology'. It is ironic that his seminal work in the field, with Loren Coleman, *The Unidentified*, is still hailed as a classic in Europe, while he himself now utterly repudiates the text and would, it seems, probably wish that he had never written it.

In *The Unidentified*, Clark and Coleman championed the views of the great psychologist Carl Jung, who studied UFOs near the end of his life. Jung never explained them to his total satisfaction, but in his book *Flying Saucers: A Modern Myth of Things Seen in the Sky*, he was the first to look towards a more humanistic approach to the evidence. Jerome Clark had embraced this and talked of the UFO as a sort of paraphysical creation of the mind that resulted from myths and archetypes from the collective unconscious describing 'other worldly' beings.

Now the wheel has turned full circle: Clark is once again convinced by the evidence that the ETH is the only viable solution. He is one of the most thorough critics of the psycho-social school of thought, and his comments have formed the basis for one of the most ambitious writing projects any ufologist has ever planned – a half-million-word, three-volume encyclopedia almost entirely written by Clark himself. With its publication in America during 1991, he expected to dispel the myth of psycho-socialism and establish the ETH once more as the best solution on offer for the evidence.

In a lecture given in New England in November 1989 Clark argued his case well. He explained that he regarded ufology as 'potentially among the most important [subjects] in the world.' He called the psycho-social perspective an 'incredibly thin edifice of theory [which] has managed to survive hurricane-force winds of evidence'.

One of the key reasons for his support of the ETH stems from the documentation of, and research into, alleged UFO crashes. The material concerned with one crash in particular, in Roswell, New Mexico, in July 1947 (see Chapter 18) leaves Clark in no doubt of its authenticity. Here claims are made that alien bodies were recovered and Clark believes these are so well established that he insists: 'If the little grey men of abduction lore are just dreams, what were they doing lying dead in the sands of New Mexico?'

His other main thrust stems from certain cases which he regards as irrefutable. In his opinion the acceptance of the reality of these events leaves little alternative but to say it is now proven that incredible machines of non-earthly design have intruded into our airspace and left proof of such visitations.

Let us examine the case cited in this lecture presented as typical proof. It occurred on 8 January 1981 at Trans-en-Provence in France and, while little known, Clark calls it 'the most important case' certainly of that decade.

The incident has just one witness, named Collini, who at 5 pm (that is, dusk or near darkness) heard a 'low whistling sound' and saw a glowing oval object descend and apparently touch down on ground at the rear of his garden. He headed towards it, but before he could get there it rose into the sky, still whistling, and disappeared. Strange marks were seen on the ground where it had 'landed' when he and his wife explored next day.

Monsieur Collini called the police, who came in daylight and took soil and plant samples from the site. A ring of two circles, one inside the other, some 6 feet (2 metres) wide, had been left. It seemed to have been indented into the alfalfa plants as if by pressure from above. The police called in GEPAN, a team of scientists which studies UFOs on behalf of the French government based at the space centre in Toulouse. Its investigation began *39 days* later, when new samples were taken. A further set to compare growth of the plants, was collected in 1983 – *two years* after the events.

Extensive research (published in a 66-page report by many scientists, psychologists, laboratories, etc.) showed beyond any reasonable doubt that the witness was sincere and faithfully reported what he believed he had seen. The ground traces were baffling, especially as the plants aged at an unnaturally rapid rate, apparently due to the intervention of some minor, localized radiation. Two laboratory analyses showed that the chlorophyll pigments 'A' and 'B' were reduced by between 30 and 50 per cent. There was also a definite statistical correlation between the damage done to the plants and their distance from the centre of the rings: the further out from the centre of the rings, the lesser the extent of this damage.

Jenny Randles met Jean-Jacques Velasco, one of the GEPAN scientists who studied the case, at a UFO gathering in Washington, DC, in 1987. There is no doubt at all that this is a fascinating example of some genuinely physical phenomenon. In Velasco's opinion, some kind of electromagnetic radiation caused the damage. The GEPAN team considered ionizing radiation but it could find no residual traces in the plants and it was unable to duplicate precisely how the biological transformation to plant life might have taken place. Nevertheless, some physical force was obviously responsible.

This case is what Jerome Clark offers as proof of the ETH, but why? He refers to the precise description of the UFO ('two saucers upside down ... with two little "legs" under it'), but we know that witnesses frequently see structure in ill-defined masses because of modern social conditioning to believe that UFOs are actual machines. Collini may have merely seen an oval mass roughly shaped like two saucers with protrusions on the underside that could readily be misinterpreted as legs.

Other factors are that it kicked up dust as it landed (which merely indicates some downward force) and that the whistling noise increased in pitch as it rose upwards (which may only mean it was rotating to create this sound and the rate of rotation increased as it climbed). Again all of this argues in favour of a real physical force but it is far from proof that it was a constructed craft, let alone a spaceship. Indeed, Collini had always believed himself that it was a novel type of military aircraft!

Again, Clark cites the puzzling ground traces as evidence, and undoubtedly something created these two concentric rings. As Clark says, the money and facilities that the GEPAN scientists could use overcomes all the charges of the sceptics who say that every UFO case would be resolved as mundane misperception if these were always provided. Here, 'all possible resources were available and the results were not ambiguous at all.'

However, can we really conclude from this, as Jerome Clark does, that 'the vehicle that landed in rural France on 8 January 1981 was not an hallucination, not a ghost, not an exotic natural phenomenon. It was – the only compellingly logical explanation informs us – an extraterrestrial spacecraft. The alternatives are more fantastic, more magical and, moreover, unnecessary?' But *are* these alternatives more fantastic?

In fact, there is extensive evidence in Britain during the same decade as this case for a rotating, glowing, whistling mass that occasionally lands and creates both flattened rings (even rings inside one another as at Trans-en-Provence) and sometimes electromagnetic disturbance to the surroundings. The so-called 'crop circles' phenomenon is not considered by anybody (even the proponents of rather exotic theories) to be the result of landing spacecraft.

Nearly 2,000 cases of such ground marks are on record during the decade in question. Views about this phenomenon range from cosmic forces to an increasingly popular hypothesis developed by atmospheric physicist Dr Terence Meaden. He proposes that a glowing plasma vortex creates the traces (see *Crop Circles: A Mystery Solved* by Randles and Fuller, Hale, 1990).

One of the consistent patterns Dr Meaden found in his analysis of the data was that this rotating atmospheric energy (which under no circumstances can be termed a 'machine'; although it is easy to see why it might get mistaken for one) does tend to appear on the slope of hills. Here the air pressure forces are most likely to generate the effect.

The Trans-en-Provence landing occurred right on the slope of a four-level terrace upon which the house of the witness was built. It is exactly the sort of location where Meaden predicts plasma vortices will form.

We must add, in conclusion, that we are not implying that this baffling case has been resolved, but that the atmospheric vortex theory has not been given an adequate airing. It may, or may not, have been a Meaden vortex that landed on this French hill slope. Indeed, Jerome Clark may ultimately be proven correct in his estimation that it was a spacecraft and this will go down in the annals as one of the best cases on record of a 'nuts and bolts' UFO. Our point is simply that this 'proof' is far from conclusive, because it proves only the presence of a real phenomenon, *not* that this phenomenon was an alien craft from outer space.

Yet Allen Hynek still thought ufology might contribute. He told *Omni*: 'It would be a tragic joke if all this money were spent searching for evidence of extraterrestrial intelligence in far places when that evidence might be under our very noses.' Perhaps by turning our attention inwards, just as Hynek suspected, we may find a quicker route to the stars; quicker than any presently available as we stand on the shores of space, gazing out at an impossibly distant cosmos.

8
George Adamski – First Contact

People have been reporting encounters with aliens for centuries. Contact with fairies and elves is an established phenomenon within the annals of folklore, and not just the stuff of children's picture books. The Industrial Revolution of the last century swept away the leprechauns and gnomes as they were now out dated. Their place was taken by the sinister pilots of incredibly secret flying machines operated by huge propellers – a fitting adaptation for the dawning technological age. Two world wars later, and we were preparing to conquer outer space. Is it mere coincidence that the 'little people' of folk tales were now being replaced by their cousins from other planets?

The archetypal 'contactee' in this new era was an American called George Adamski. Born in 1891, his family emigrated to America when he was just 2 years old. Adamski became something of a self-made man, dabbling in philosophy and astronomy, and moving through a series of menial jobs, mostly at the bottom end of the catering market. Fame and fortune smiled upon him after he made some quite extraordinary claims. Adamski said he had met aliens and had ridden in their space vehicles. What is more, he produced photographic evidence to back up his story and the testimony of many other witnesses.

His adventures were chronicled in several books: *Flying Saucers Have Landed*, with Desmond Leslie, *Inside the Spaceships*, *Flying Saucers Farewell* and the privately published, *Cosmic Philosophy*. Although in the first person, all the accounts were ghost written.

The first encounter took place in the Californian desert just after noon on Thursday, 20 November 1952. Adamski was accompanied by his 'secretary', Mrs Lucy McGinnis, and by Mrs Alice Wells, Mr and Mrs Bailey and Dr and Mrs Williamson. He had already taken a number of photographs depicting unknown objects, starting back in 1948. During that post-war period, America was escalating towards fever pitch at the growing number of UFO sightings – unknown aerial craft which the government could not, or would not, explain adequately. Against this background, Adamski's party drove out into the desert to have a picnic.

After changing a flat tyre, they stopped for lunch. Williamson, a geologist, busied himself examining some rocks. Suddenly their attention was drawn to a gigantic, cigar-shaped craft hovering, silently, above

The most famous contactee of all –
George Adamski. His claims of
meeting humanoid aliens sounded like
post-war fairy tales, but there are
aspects of his case which still beg
explanation.

some mountains. Something made Adamski think that a landing would occur and that the ship had come looking for him!

With two of his friends, he loaded his camera gear into the car and set off to a spot almost directly beneath the giant object. They stopped about half a mile (one kilometre) from the rest of the party. Adamski asked his companions to return and take note of anything they saw. As the car sped away, the cigar shot upwards out of sight as military planes roared overhead.

While Adamski struggled to set up his equipment, he was distracted by a bright flash in the sky. He watched a small craft drift through a saddle between two peaks and land about half a mile away just over a ridge. A few minutes later, a figure appeared and beckoned Adamski to come over. Apprehensively, he did as he was bidden, noticing the 'man' wore a one-piece jump suit and had unfashionably long, blond hair which blew in the wind. Then it hit him. He was in the presence of a man from another world.

Telepathically, the man said that he was from Venus and that his race was concerned with the radiation from atomic bombs reaching into space and harming other planets. Adamski was also told that the small saucer-shaped 'scout' ship had come down from the cigar-shaped 'mother ship'. Aliens were also visiting our world from other parts of the solar system and beyond.

This was only the first of a long string of incidents which elevated George Adamski to the status of a guru and caused a controversy which still rages.

Following the desert episode, Adamski described how he experienced a strong telepathic desire to visit Los Angeles, where he was approached by two of the humanoid aliens in a hotel lounge. They took him into the desert, where he was transported to a mother ship. Here he was introduced to other aliens and grand philosophic debates took place. Adamski's other claims included a trip around the dark side of the moon and a visit to Saturn – also populated by humanoids!

With the benefit of hindsight, it is very easy to sweep the whole affair aside as nothing more than imaginative tale-telling and gullibility on the part of Adamski's followers. Yet, although his first book told a sensational story, it was logical and, for the times, not beyond the realms of credibility. Contact with aliens from Venus was accepted in the 1950s as an explanation for the rush of UFO reports. It must not be forgotten that the other witnesses to the desert encounter, who had observed events through binoculars, signed sworn affidavits, although, according to writer Frank Edwards, some of them later changed their minds. Subsequent stories of further meetings and flights into space did seem to detract largely from the initial event.

Some ufologists believe that Adamski's timely arrival on the scene did irreparable damage to the subject. Until then the phenomenon had seemed worthy of scientific consideration. His later claims sent the credibility horizons of many people zooming into hyperspace! The episodes seemed straight out of the monochrome science-fiction films of the period, except that much of the action had been sacrificed by some very naïve philosophy on the part of the 'space brothers'.

It was obvious, whether Adamski realized it or not, that the ideals of this cosy galactic club were a reflection of his own philosophies and fear for the future of humankind. Indeed, the second book was little more than an extrapolation of a short story written by him in 1949. *A Trip to the Moon, Mars and Venus*.

This is what Frank Edwards had to say on the matter:

His efforts did not attract many customers but it did attract the attention of a lady writer who saw gold in them there space ships. She made a deal with George to re-write his epic; she was to furnish the skilled writing and he was to furnish the photographs of the space ships.

Edwards claims to have seen the manuscript, with 'a clutch of the crudest UFO photographs I have seen in years.' Respected researcher, Timothy Good, in his excellent book *George Adamski: The Untold Story*, remarks on this assertion that Adamski merely adapted this fiction to his later UFO books: 'Although there are indeed far too many similarities to be purely coincidental, it is true that Adamski's fiction was wildly off the mark in other sections by comparison with some of his later claims.'

What really attracted so much attention to the story were the many photographs of cigar- and saucer-shaped craft Adamski produced.

Without these, his lurid accounts would probably have died a death early on. Pictures of the 'scout ship', featured heavily in *Flying Saucers Have Landed*, have come in for a lot of debate. The object has been compared to a lampshade, chicken feeder and numerous other items, including a part off a vacuum cleaner. Recently, someone has produced a vintage overhead snooker lamp which bears a strong resemblance to the spaceship, however, none of these objects match exactly the thing which Adamski photographed.

Was Adamski's saucer made in Britain, Wigan, Lancashire? In 1975, the mystery seemed to have been solved when the top of a bottle-cooling machine was found to match the photographs almost exactly. However, when the designer was traced, he proved that the top has been made in 1959, seven years after the pictures were released. Apparently the engineer had been a fan of Adamski.

Another picture in the book, showing a blurred saucer passing low over some trees, is credited to Sergeant Jerrold Baker and taken on 3 December 1952. A sworn statement by this man was published in Gray Barker's *Book of Adamski*.

On Saturday morning Adamski called me and said he saw what he thought to be a saucer coming from the coast. I hurried up the hill and stood by a large tree. Suddenly I saw a circular object skim over the treetops from the general area where Adamski was located. It was a flying saucer -- of that I was sure. I seriously thought it was going to land in the small clearing because of its extremely low altitude. I waited momentarily, mostly because of shock I guess, as it continued coming closer.

It then hung in the air not over twelve feet (3.7 metres) high at the most, and about twenty-five feet (7.6 metres) from where I was standing. I quickly snapped a picture and as I did it tilted slightly and zoomed upwards over the tree faster than anyone can imagine!

Eighteen months later, in a letter dated 29 June 1954, Baker repudiated this, saying he did not take the picture credited to him. In view of the lapsed time between publication of the photograph, and then denial, and bearing in mind Baker's sensitive military status, Tim Good wonders if he was put under pressure to deny this sworn statement by the US Air Force.

If the entire affair were nothing more than science fiction and a handful of camera tricks, it had some remarkable ramifications and a host of willing confederates. Why did the 'scout ship', with its unique three-ball underside 'landing gear', start turning up elsewhere?

During February 1954, at Coniston, in the heart of England's Lake District, two boys went out for a walk. One of them, the son of a doctor, carried a camera. Just over the crest of a hill, they came across a strange hovering machine. A picture was hurriedly taken and the object departed. The resultant image is blurred, but clear enough to show the outline of an Adamski saucer.

Leonard Cramp, an aeronautical designer and engineer, in his book *Space, Gravity and the Flying Saucer*, scientifically compared one of Adamski's pictures with that taken by 14-year-old Stephen Darbishire. Using a system called orthographic projection, he proved that both objects were proportionally the same, even though they had been photographed at different angles. Investigator Harry Hudson, who recently spoke to Darbishire, told us that the man still sticks to his story.

We investigated a case in Warrington, Cheshire, where an entire family observed a strange object passing over their house. This too was saucer-shaped with three balls in a triangular configuration underneath. The family had never heard of Adamski, never mind concocted a hoax sighting to link in with something which had occurred 30 years before. The most recent case we are aware of happened in August 1987, in Rochdale, Lancashire. The witness, an artist, observed several of the objects passing over his garden. Investigator Philip Mantle of the Independent UFO Network, could find no easy explanation.

Adamski was heavily criticized for claiming to have been taken on a flight past the dark side of the moon. He described mountain peaks covered in snow and heavily wooded valleys! This has since proved to be nonsense, along with the claims that humanoid civilizations existed on our planetary neighbours. But not all of his 'observations' were wrong. On page 67 of *Inside the Space Ships*, he describes his first alleged trip into space:

> I was amazed to see that the background of space is totally dark. Yet there were manifestations taking place all around us, as though billions upon billions of fire-flies were flickering everywhere, as fire-flies do. However, these were of many colours, a gigantic celestial firework display that was beautiful to the point of being awesome.

On 20 February 1962, astronaut John Glenn, orbiting earth in a Mercury space capsule, described the following: 'a lot of the little things I thought were stars were actually a bright yellowish green about the size and intensity as looking at a fire-fly on a real dark night . . . there were literally thousands of them'.

Further corroboration was given by Russian cosmonauts Vladimir Komarov and Konstantin Foektistov in Voskhod 1 during 12 October 1964. Apparently this phenomenon is caused by billions of reflective dust particles. Adamski's description appeared in print several years before this phenomenon was observed by our astronauts – remarkably accurate guesswork, or did George Adamski actually travel into space aboard an alien spacecraft?

The hoax theory took another twist with the publication of *Flying Saucers: An Analysis of the Air Force Project Blue Book Special Report No. 14*, by Leon Davidson, a chemical engineer. He suggested Adamski was the naïve victim of a gigantic fraud perpetrated by government agents. The spacecraft Adamski claims to have travelled in, Davidson believed, was a man-made structure, showing filmed vistas of space. To support his theory, he pointed out that Adamski was encouraged in his extraterrestrial search by four government scientists. This would be part of an international conspiracy to unite the peoples of earth against a common enemy.

We do not find this credible. Adamski's aliens were jolly fellows, more like saviours than enemies. However, this theory could explain his fall from grace during the later part of his life. If he were being used by the CIA, it would have been easy to bring about a character assassination once his usefulness had ended. A more tenable reason for CIA involvement would be to defuse serious interest in the UFO subject through the downfall of Adamski as his stories become more outrageous.

People who met Adamski personally were impressed by his apparent sincerity and honesty. In his book *UFO – Flying Saucers Over Britain*, science journalist Robert Chapman, an admitted sceptic of the grey-haired American, recounts such a meeting.

Adamski responded easily and without hesitation in support of his remarkable claim. It had happened and that was that. If anyone believed him he was glad; if they did not it was too bad but what could he do about it? Long before I left him I knew I was beaten as far as tripping him into any incautious admission was concerned. Adamski was so damnably normal and this was the overall impression I carried away. He believed he had made contact with a man from Venus, and he did not see why anyone should disbelieve him. I told myself that if he was deluded he was the most lucid and intelligent man I had met.

Sceptical ufologist Steuart Campbell made this remark in the April 1983 edition of *The Probe Report*:

Certainly when I met him in April 1959, on one of his European tours, he was an accomplished and experienced raconteur, with an unshakable story. However, confidence should not be mistaken for veracity.

But would a trickster be granted an audience with the Pope? According to the late Lou Zinsstag, co-author of *George Adamski – The Untold Story*, she and Belgian acolyte, May Morlet, accompanied Adamski to Rome, where, he claimed, he had an appointment with Pope John XXIII. At 11 am on 31 May 1963, they watched Adamski enter the Vatican, not by the main gate posted with Swiss Guards, but discreetly by a side door. He returned an hour later 'grinning like a monkey' and bearing a gold

Still from an 8 mm ciné film taken by George Adamski in the presence of Madeleine Rodeffer, of Silver Spring, Maryland, on 26 February 1965 – just months before Adamski's death. The craft allegedly hovered 70 to 90 feet (21 to 27 metres) from the witnesses. William T. Sherwood, an optical physicist, thought the film was genuine. The couple had been warned of the imminent saucer sighting by three 'men' in a car which pulled up outside the house. (*Madeleine Rodeffer*)

ecumenical coin presented to him by the Pope. Unfortunately, the nature of this clandestine visit meant it was not recorded in the daily visitors' list. Two days later the Pope died.

By 1965 Adamski's popularity was on the decline. Even some of his most ardent followers had deserted him. This was partly due to his immersion into something bordering on spiritualism, and an even wilder claim that he had visited the planet Saturn. Then he produced a piece of ciné film apparently showing a saucer, allegedly taken on 26 February of that year.

He had been the guest of Nelson and Madeleine Rodeffer at their home in Silver Spring, Maryland. Adamski informed Madeleine that one of the aliens had warned him to have a camera at the ready as one of their craft would be flying by. As Madeleine had received a new 8 mm movie camera for Christmas, and she was not used to operating it, Adamski helped her load it with film.

In mid-afternoon they caught sight of something hovering over nearby trees. At the same time, a grey car drew up and three men climbed out, apparently extraterrestrials. They shouted, 'Get your cameras – they're here!'

Madeleine panicked and found she could not operate the camera, so she handed it to Adamski, who took the film. The object did a series of manoeuvres, changed colour several times and even altered its shape. All this apparently was captured on the film, but after its return from processing, several frames were missing and some fake footage had been added, claimed Adamski.

The resultant collage came under a lot of fire. But in a letter addressed to Tim Good, the then editors of *Flying Saucer Review*, Charles Bowen and Gordon Creighton, remarked: 'I don't think either Charles or I ever thought the pictures were fakes. I simply think that Bowen and I felt they looked like the usual "transmogrifications" . . . fakes by "them", not by humans.'

Perhaps a lot of the scepticism evolved from the way in which Adamski's claims were presented in the three main books. For instance, many people have remarked on the clichéd names of the aliens: Orthon, Firkon, Ilmuth and Kalna. Apparently Adamski claimed that the extraterrestrials did not have names, and that these were a literary device created by ghost-writer Charlotte Blodget to help the flow of the story. One wonders how many more facets of his claims were presented in a totally different way to match the expectations of the times, and make the books more commercially viable.

Perhaps the tales of friendly aliens and jolly jaunts around space were nothing but a façade for something darker and sinister. Alternatively, Adamski may have seen the growing interest in flying-saucer sightings as the perfect vehicle to bring to the attention of his fellow men the dangers in which they were placing the planet earth – dangers of which, 40 years on, show signs of coming to fruition.

9
Life on Mars

Prior to our exploration of the solar system in the 1960s and 1970s, Mars was considered by many to be a safe bet for extraterrestrial life. Ever since Schiaparelli noticed a series of lines crossing the red planet in 1877, and referred to them as *canali* – meaning channels, but widely mistranslated as 'canals' – the dreamers have been dreaming.

Science-fiction authors, like H. G. Wells, Edgar Rice Burroughs and Ray Bradbury, took Mars into their bosom and endowed it with tales of men and monsters, and aliens every bit as sinister and exotic as the planet itself. But the fictions were only an externalization of the hopes, fears and dreams of human beings. Could there really be an alien civilization on Mars? Many people thought so; Hélène Smith knew so.

Smith was a pseudonym for Catherine Elise Muller. In the 1890s she believed she was in psychic contact with extraterrestrials on Mars. Her detailed and complex revelations attracted the attention of the prominent Swiss psychologist Théodore Flournoy. Even though he befriended the young woman for many years during his studies, Flournoy did not compromise his stance that this psychic information should not be taken at face value. It says a lot for Catherine that she accepted his open scepticism, while hanging on to her beliefs.

Catherine believed she was 'born for better things', and even questioned her parents whether she was truly their child. Flournoy speculated that the whole affair was a subconscious revolt against drab, everyday life, externalized in the form of exotic fantasies. But, as the psychologist was to learn, the answer was not that simple. Catherine was bright and intelligent, and held down a responsible office job. Yet she admitted to psychic experiences from childhood. On top of this, her mother had received a vision of an angel when Catherine's younger sister died. There seemed to be a hereditary disposition towards psychic visions in the family. Théodore Flournoy did not view Catherine's communications as pathological, but as 'rare' and 'exceptional'.

She produced details about the Martians – their houses, society and flora; but the narrative, although detailed, was often incoherent. Flournoy's observations led him to conclude Catherine was living a dual existence: a part of her mind was living the Martian existence every minute of the night and day. When she 'switched' over into the Martian mode, it was like dipping into the pages of a book, jumping on to a moving escalator, which did not stop moving just because the shopper was not there. This duality was illustrated in an incident which took place during the night of 5 September 1896.

Catherine was awoken in her room by a high wind which rocked the house. Afraid that some flowers set out on the sill of her window would be blown away and destroyed, she climbed out of bed and brought them in. Afterwards she sat on the edge of the bed and found it transformed itself into a bench on the edge of a pinkish-blue lake. A bridge, edged with transparent yellow tubes, spanned the lake. This was Mars, and around her were the Martians.

The landscape was crowded with exotic 'people'. One of them carried a small device similar to a lamp, which allowed him to fly. The experience lasted 25 minutes, and during this time Catherine was fully aware of her surroundings, convinced she was wide awake. There were many experiences like this which added to the information on this alien world. More remarkable was the Martian language she began using.

This was not gibberish, Flournoy discovered. Catherine would use the same word, weeks apart, with the same meaning. It was a fully structured language with coherent syntax. In many ways it was similar to French. Flournoy wondered if she had subconsciously created this language and committed it to memory for use as easily as her own. This in itself was a remarkable feat. However, Catherine would have none of Flournoy's 'reasoning'. As far as she was concerned, she was in contact with an alien culture on the planet Mars.

Author 'Cedric Allingham' went one step further than Catherine Elise Muller when he claimed in his book *Flying Saucer From Mars* that he had actually met a Martian.

According to his own story in his book, Allingham motored to Scotland from London for a holiday to study bird life and do some hiking. On 18 February 1954, while strolling along the coast between Lossiemouth and Buckie, he heard the sound of an engine and became aware of a 'swishing' overhead. High in the sky he saw a dark speck which binoculars resolved as being saucer-shaped.

Unperturbed, Allingham stopped for a picnic lunch before continuing his walk. After sighting the object a second time, it returned from across the sea towards him. In his account, Allingham describes the noise of the engines and how he took some photographs as the thing landed close by.

As he stepped forward to observe more detail, a sliding panel near the base opened and a figure leapt out. The 'spaceman' was humanoid, about 32 earth years old, 6 foot (nearly 2 metres) tall and wearing a one-piece suit. Mounted on his back was some sort of breathing apparatus.

The 'flying saucer', obviously based on George Adamski's Venusian scout craft, allegedly photographed by Cedric Allingham in Scotland. Note what appears to be a thin wire or thread attached to the dome. (*Cedric Allingham*)

With the aid of sign language and sketches, Allingham found out the being came from Mars. After an exchange of general information, including several revelations regarding the 'canals', the spacecraft departed.

During the walk back, Allingham realized there was no one to support his incredible story. Then on the road to Lossiemouth, he found a fisherman called James Duncan who excitedly confirmed the sighting. He had observed the entire episode from a hilltop. The Scotsman signed a sworn statement, and armed with this, Cedric Allingham returned to London to begin work on a book.

Flying Saucer From Mars was published in 1954 by Frederick Muller. It caused quite a stir and sold thousands of copies. But there were sceptics, such as journalist and science writer Robert Chapman, who thought the alleged conversation between Allingham and the alien was too terrestrial in terminology. Also he thought the photographs were blurred and wondered why were there no clearer shots?

The attitude of the publishers also fuelled his suspicions. When reporters attempted to interview the author, they met with a wall of silence. This seemed strange. Publishers and writers of bizarre stories are usually happy to talk to the media and gain as much publicity as possible. Frederick Muller claimed Allingham had gone to America in search of famous contactee George Adamski. Then they told Chapman he had gone to Switzerland to convalesce after a serious illness. Finally it was announced that he had died.

Attempts at tracing 'James Duncan' proved just as fruitless. However, we have documentation from one man claiming to have met Duncan in 1968 and 1973. According to this lucid account, the Scotsman *had* witnessed something, but it was rather different and less spectacular than Allingham's book alleges. Duncan even refused to read *Flying Saucer from Mars* because of the alleged distortions. Were both characters nothing more than a literary device in a thinly disguised novel? Yet at the time of publication, a man claiming to be Cedric Allingham lectured before an audience of flying-saucer devotees in Kent. So what was the truth?

The hue and cry eventually died away. Robert Chapman, in his 1969 book *UFO: Flying Saucers over Britain* concluded the affair 'was probably the biggest UFO leg-pull ever perpetrated in Britain.' Others said similar things in later books. That seemed to be the end of the subject, until a decade afterwards when two ufologists decided to get to the bottom of the mystery once and for all.

Christopher Allan and Steuart Campbell carried out some enquiries and reached the astonishing conclusion that the culprit could be famous TV astronomer Patrick Moore.

Allan and Campbell were able to use a computer at Edinburgh University to analyse the text with a programme based on 'stylometry', which could determine whether *Flying Saucer From Mars* had similar word patterns to Moore's other books. The results, even in extended searches, were inconclusive. Stylometry had failed to nail the culprit.

Then, the two researchers received from Muller's the name and address of someone whom they said was connected with the long out-of-

Cover of *Flying Saucer From Mars*. The 'facts' presented by Cedric Allingham did not check out. (*Cedric Allingham*)

Sceptical ufologist Steuart Campbell helped uncover the hoaxer behind the 1950s book *Flying Saucer From Mars*. (*Peter A. Hough*)

print book. His name was Peter Davies, a journalist who acknowledged that he was an old friend of Moore. It was an interview with this man that finally seemed to clear up the mystery. He admitted the book was a hoax, but that his part had been to re-write the manuscript and disguise the style of the original author – hence the failure of stylometry.

Davies said he had split the royalties with his co-author, but would not say who that was. He also agreed that the picture on the cover depicting Allingham, was himself, heavily disguised. 'That's the best picture in the book,' he joked, 'and even that one's a fake!'

Patrick Moore has remained adamant that he was not the author of the Allingham book.

Allan and Campbell published their findings in the July 1986 issue of the journal *Magonia*. Soon afterwards we mentioned in *Exploring the Supernatural* magazine that it was now widely believed that the Allingham story was a hoax, but we decided not to accuse anybody. We feel readers can only make a balanced judgement from considering the full details.

The Allingham and Adamski books came at a time when we did not know for certain whether life existed on Mars. Since then, despite the fact that Mariner and Viking spacecraft have demonstrated that Mars does not have an earth-type environment, and therefore is not conducive to humanoid life forms, the dream refuses to go away. Although some of the older mysteries have now been solved, they have been replaced by many new ones.

The atmosphere on Mars is too thin to support life, yet an atmosphere does exist. Permafrost (possibly of carbon dioxide) has been discovered at the poles, and the romantic-sounding *canali* may turn out to be ancient dried-up river beds, if some geologists researching the data are correct. Although much of Mars is cratered, other parts are not. Raging dust-storms can spring up at speeds in excess of 200 mph (300 kph), and the planet boasts the largest volcanic pile ever discovered: *Vide Olympus Mons*. It is 15½ miles (25 kilometres) high and 435 miles (700 kilometres) wide!

All these factors have caused many people to speculate that the Mars we see today may be the withered remains of a once-thriving world not unlike our own. And if that is the case, there might indeed be the traces of life of some sort waiting to be discovered.

The Viking 1 mission, which left the Kennedy Space Center on 20 August 1975, brought the Red Planet amazingly to life with clear colour pictures of the surface. Eleven months after leaving the earth, the Viking landers touched down at separate locations on Mars. The surface, a mixture of thickish-red soil, scattered with rocks, bore a strong resemblance to desert regions on earth. There were pink clouds too; in fact all the basic building blocks of life exist in our sister world: carbon, oxygen, hydrogen and nitrogen. While life was not detected, several of the pictures sent back to earth suggest what some believe are artefacts from an ancient civilization!

The first things noticed were rocks with letters or numbers on them. On one there is a B or an 8. On others the letters O, S and T. Although this is intriguing, if you look long and hard enough, scratches and marks would probably be found that look like the entire alphabet. But what would characters from the English alphabet be doing on Mars? More surprises were to come.

From its orbit of 1,162 miles (1,870 kilometres) above the Cydonia Plains, Viking captured a sphinx-like face, almost 1 mile (1.6 kilometres) across, and a pyramid in what some researchers have described as the ruins of a city! Giant monuments constructed by Martian engineers?

The official NASA view is that the effect is caused by a combination of light, shadow and wishfull thinking. We tend to seek patterns in inanimate objects, as demonstrated by fairies or elves discovered sitting in trees on photographs when all that is visible is produced by sunlight shining randomly through leaves and branches.

But not all scientists agree. Vincent DiPietro and Gregory Molenaar, former computer scientists at NASA's Goddard Space Flight Center in Maryland, and computer expert Richard Hoagland of Oakland, California, applied their considerable technical skills to the sphinx's face, and uncovered details on the shadowy side, plus an eye ball within the dark eye socket. According to Hoagland, who later wrote *The Monuments of Mars*: 'This processing effectively eliminates the idea that the face is a trick of the light. It clearly points to it being the result of artificial construction.'

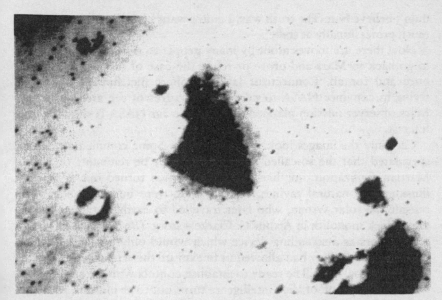

The remarkable 'face' on the surface of Mars taken from Viking 1. Is this really an alien construction or an accident of light and shade? Compare it with the rock simulcra on the Sedona photograph on page 81. (*NASA*)

Crucial in the argument of these researchers is their denial of NASA's claim that later orbital pictures of the same region do not show the face, thus proving it to be an accident of light and shade. They point out that photographs taken on both 20 July 1976 (NASA reference 35A72) and 24 August 1976 (reference 70A13) both show the image, despite being taken under differing conditions and at other angles.

They also have some powerful support in the form of Dr Brian O'Leary, a physicist and astronomer who trained as an astronaut to go on NASA's manned Mars landings before budget cuts officially scrapped them. He then moved to Cornell University and started work on the detailed analysis of planetary missions, searching for evidence of the history of our neighbouring planets and possible signs of life.

O'Leary is not a scientist who rejects unconventional evidence out of hand. He has explored the phenomenon of channelling and has discussed UFO abductee-type experiences with one of America's best-known witnesses. His book on these phenomena, *Exploring Inner and Outer Space* is candid and frank. According to a report by Sean Devney (*UFO Universe*, May 1990), he also changed his mind on the photographs of the face on Mars after initially dismissing them.

After hearing Richard Hoagland at a Colorado symposium on Mars in 1984, he re-studied the pictures and said: 'At the time I listened, but didn't believe him. The truth was, I didn't want to believe him. I had too much professionally at stake'.

Now there are moves afoot by many people to try to persuade NASA to go back to Mars and prove or refute the case of the Martian sphinx once and for all. Connecticut lawyer Robert Bletchman is currently trying to convince NASA to take more pictures of the area during the Mars observer mission planned for 1992. So far NASA is showing little interest.

Certainly the images look very interesting. Some commentators have speculated that the so-called artefacts may not be remnants of a dead Martian civilization, nor like the canals, which turned out to be both illusions and natural ravines. Instead, they were built by aliens from outside the solar system, who later travelled to earth. Perhaps, like the great black monolith in Arthur C. Clarke's story *The Sentinel*, they were set on Mars as a signalling device which would only be activated once our space technology had allowed us to explore the artefacts thoroughly. Then, perhaps, we will be ready to establish contact with the architects!

But if this 'proof' of alien intelligence turns out to be illusion, and Mars remains depressingly empty, then the myths will no doubt still simmer on, because such myths are necessary and feed upon our need to find some company in space.

10
Channelling and the Inner-Space Connection

The rationalist British UFO researcher Hilary Evans described the Hélène Smith story in the September-October 1986 edition of the *International UFO Reporter* in this manner: '[This] is probably the most revealing case in the literature of alleged communications between human beings and extra-terrestrials.'

This fascinating case either shows the amazing hidden capabilities of the human mind (particularly with certain gifted people) or else it is a genuine example of Martians choosing to project their visions into the mind of one Swiss woman. It is easy to understand why this case was of great interest to both the mystics and the more avant-garde school of psychology (for example, Jung, who himself reported on similar incidents from the annals).

While Evans reaches the inevitable conclusion that this is just an illustration of how 'the unconscious mind, in creating psychic fantasies, draws on whatever it finds suitable for its purpose', the case also offers us surprising insights into the latest phenomenon of the 1990s – a century after the experiences of Hélène Smith. This new craze is called 'channelling'. A channeller is, in truth, little more than a spiritual medium who relays psychic messages from outer space rather than from 'the other side'. The tradition of a 'spirit guide', who is a North American Indian with a name like 'Running Deer', has been replaced by an entourage of aliens with scarcely more authentic sounding appellations like 'Zookan' and 'Maximus'.

As the case of Hélène Smith amply demonstrates, there is nothing new in heaven, earth or other solar systems.

A typical modern tale that mirrors Hélène Smith's almost exactly, concerns a young woman. She too is highly intelligent and successful, not the social loser who critics tend to presume channellers must be. 'Caron' had a history throughout childhood of strange events – seeing ghosts, weird creatures, visions, premonitions, etc. Then her experiences extended into the realms of the UFO phenomenon. It became a regular outlet and she began to undergo contacts and visionary trips to unknown worlds. She described the landscape and the fauna of these and even learnt the names of aliens and creatures of many other planets. The contacts took place during the 1970s and 1980s.

Researcher Hilary Evans believes
social and psychological factors are
important keys to understanding
claims of contact with aliens. He
believes the Hélène Smith case is a
prime example of his hypothesis.
(*Peter A. Hough*)

Overall, the similarity between the two cases is extraordinary, yet
Caron was never referred to a psychologist. She was never considered in
any sense unstable and she had no problem in accommodating these
phenomena into her daily life. Most of her friends probably know
nothing about her 'other' life. Nor is Caron's case by any means unique.
We have investigated about a dozen similar ones. There are many more
on record (Mantle, Potter and Walmsley summarize four in an article in
UFO Universe, May 1990). In America especially, the community of
'mental contactees' has grown to prodigious levels, particularly in
California and the Arizona desert. There is even an entire town of New
Age people in Arizona!

Jenny spent some time as an observer at a New Age symposium in
Arizona, which added to the information we have obtained from our
British contacts. Despite the apparent absurdity of these people's claims
there is little doubt about the sincerity of most of them. They are usually
clever and intelligent, and possess a viewpoint that seems eccentric only
when viewed against the norm. The New Age (with its philosophy that
the earth is about to go through a spiritual revolution using ESP and
consciousness-expansion) warmly accommodates this. Many of the
channellers are extremely gifted at some form of creativity of visualiza-
tion. Recently, former TV presenter and politician, David Icke, has
popularized the phenomenon through his own claims. The New Age
seminars tend to be filled with examples of their outstanding artwork,
jewellery design and poetry. Channellers also have what might be
variously described as 'wonderful imaginations' or 'exceptional visual
acuity'.

Of course, this does not really answer the basic question about
whether they are simply gifted people who can 'tune in' to things that are

blocked out from the rest of us, or whether they have such well-developed creative and visual abilities that at times the products of their inner selves become impossible to distinguish from what we call real-world experiences.

Indeed, some even suggest there might be a third alternative. Since all of our reality is a product of our senses, perhaps we determine what is real by virtue of majority rule. In other words, our universe might be fundamentally different from the one in which the channellers live.

Is there any difference between the channellers and the contactees who dominated the UFO scene during the 1950s? Again, there are some who say yes and others who think the two are different expressions of the same state.

We can see the similarities when we look at the names of the aliens or their home planets as reported in contact stories – Aura Rhanes, Clarion, Ashtar, for instance. The contactee messages also usually comprised friendly warnings about the mess we were making of our environment and the need to join some sort of universal brotherhood of peace, although there was rarely any agreement between individual cases, as if each contactee chanced upon an entirely different alien race during its one and only mission to earth!

However, there were also differences. Most of the 1950's contactees claimed physical meetings or trips in flying saucers. This was why they

This New Age community has been set up in the red rock country around Sedona, Arizona. Here psychics, channellers and other esoteric believers live together. Note the human face on the rock to the left. This is simulcra, an accident of erosion and lighting, or – as some believe – an alien artefact like the face on Mars. (*Jenny Randles*)

The amazing popularity of the New Age channeller movement is shown by this packed audience of several hundred who paid to hear leading figures present their views at a Fall 1989 conference in Phoenix, Arizona. (*Jenny Randles*)

became popular celebrities. They offered fuzzy photographs as proof of this. The majority were men with some sort of vague technical background, whereas channellers, like mediums, are probably more likely to be women with far less of a scientific standing. One of the most common consequences of a contactee meeting with aliens was the publication of the obligatory book and the ego-boosting lecture tour. None of these things are found within the community of channellers, at least not in such an overt fashion.

Perhaps a more thorough study will come to a conclusion about their similarity or divergence. That said, possibly because of our lack of historical perspective through living alongside them, there does seem something superficially more plausible and persuasive about the channellers, or 'space mediums'. However, that is not the same as saying that the messages they convey convince us of their other worldly origins. UFO historian Dr Jacques Vallée said, 'Just because a message comes from heaven doesn't mean it isn't stupid'. By which he implied that all such messages could be absolute nonsense even though the medium through which they pass is totally sincere.

Before the emergence of modern channelling there was another phenomenon that is related, yet interestingly different. This was instigated by what some call the Extraterrestrial Travel Agents – strangers who appeared in a community, staged lectures and encouraged 'followers' to give up their lifestyles and prepare to go with them on a

UFO, always due shortly, which would whisk them away to peace and safety on another planet as the earth faced destruction.

The origin of this movement goes back to about 1975 when a man and a woman (usually given the names 'Him and Her' or 'The Two') went on a pilgrimage around the western states of America. Posters for their lectures suggested that this was 'not a discussion of UFO sightings or phenomena' nor was it 'a religious or philosophical organization recruiting membership'. Supposedly, they were 'sent from the level above human' and wanted to explain to willing people 'how the transition from the human level to the next level is accomplished'. Once achieved, they could all 'return to that level in a space ship'.

Nuclear holocausts and oil crises helped to generate anxiety about the future of the earth. Even today there are 'survivalist' communities of people so afraid of a coming catastrophe that they have given up all their possessions to live in a remote fortress hideout in the hope of escaping calamity.

Similar fears about the earth's future seem to have fuelled the success of 'Him and Her' and other prophets of doom who followed in their wake. While people did change their lifestyles (there were stories of houses being given away or sold for just $5), as far as we know the spaceships never came and no one made the transition.

It would be simple to dismiss all of this as a con trick, but there seems scant evidence for it. In any case, even if any of these alarmists were outright phonies, the victims conned themselves into believing that the grass would be greener on the other side of the cosmos.

In the late 1970s another couple, Michael and Aurora El-Legion, went on an international tour to advise the world that there were 'star people' on earth. When they visited Britain, they were treated during TV debates as representatives of the UFO movement, which caused much anger among serious ufologists. At one location there was a major Church protest outside the lecture hall, with attempts to have the sessions banned.

The El-Legions appear to have been sincere forerunners of the channeller movement and still make appearances at New Age gatherings in America. They were important for popularizing the idea that the earth was facing dangerous times and that the only way to avoid them was to turn to the aliens for help.

Of course, this is a very interesting premise which a growing number of people still seem willing to embrace. Today, fears that the earth will be destroyed by nuclear war have been shunted to one side and replaced by those concerned with pollution and the greenhouse effect.

Quite recently a new ambassador from France has been doing the rounds. Looking like the clichéd Indian mystic in flowing robes, he has proposed a unique idea to establish an intergalactic embassy where the spaceships can land and take up residence. His blend of cosmic philosophy strikes a chord within the younger elements of society. One British TV show welcomed him, but insisted upon a teenage ufologist standing by his side because the message which he wanted to convey had to reach that audience. This TV programme then devoted an hour to the building of a mock alien embassy (including palm trees!) in the studios.

One is left bemused by what this is supposed to have achieved, although it does illustrate the dramatic pull such ideas have on the young within society. UFOs, as such, are boring old hat; this is more fun.

This pre-channeller phase was cleverly portrayed in the film *The Mysterious Two* (1982). It starred Priscilla Pointer and John Forsythe as the lecturing couple, which is rather ironic as Forsythe later played mogul Blake Carrington in the TV series *Dynasty* when in 1987 it featured the infamous scenes where his screen daughter, Fallon, was abducted in a pyrotechnic extravaganza based on the real-life alien kidnap cases then much in the news.

The Mysterious Two was a well-constructed and understated film. Reportedly entitled *Follow Me If You Dare* until almost the day of release, it was fictional, but obviously leant very heavily on the actual cases of couples such as 'Him and Her' (using character names 'He' and 'She' instead). It sought neither to explain away nor justify what happened, but it had to include the token UFO, although this was discreetly kept to a few pulsing lights which the entranced locals could ignore. What the film did display effectively is how it was possible for two sincere strangers, with frankly incredible stories, to have led astray innocently hundreds of people who lost everything in the pursuit of what for most of them was a dream without a happy ending.

The channellers of today come in the wake of these uneasy precedents, which explains why some aspects of our society (notably the New Age community) accept them warmly while the majority are more ambivalent, either considering what they offer as self-delusion or (if they are even less charitable about such matters – and many are) regarding them as people who are simply out to make a fast buck at the expense of a gullible public.

This last view would not be a fair description of most channellers, at least not of the ones we have met. Whatever the truth about the origins of their messages, most convey them with apparent sincerity.

Diane Tessman is probably the archetypal New Age channeller. With the looks of a fashion model (many channellers are from the 'beautiful people' set) and the convincing gift of powerful communication, she claims to receive messages in her mind that come from another being.

Then, just as with certain types of spiritualists (known as 'direct-voice mediums'), she allows the alien to communicate with the outside world by speaking through her larynx. Some talk as they normally would, others adopt different tones and timbres. All are claiming to be nothing more than a channel for another living creature from a distant world.

This can be faked, as members of the sceptics community have been quick to point out. Magician James Randi trained one lucky youth to be a channeller in a couple of weeks. He just happened to be in Randi's house when the arch-debunker was asked to go to Australia to prove his thesis. He did, and the youngster filled the Sydney Opera House with his channelling act. Randi says members of the crowd offered huge sums for a 'crystal' the channeller used to aid his 'communications'. In reality, it was just a lump of rock picked up in the car-park on the way in! Most channellers are not charlatans, so Randi only proved how easily people can be taken in – if they want to be taken in.

Diane Tessman appears to be very sincere. Her contact is named Tibus. She first established contact in a childhood experience relived under hypnosis with the help of University of Wyoming psychologist and Professor Emeritus, Dr Leo Sprinkle. Her story has numerous parallels with that of the UFO abductee – a person who claims no regular contact with aliens, just the occasional (or even singular) and often very traumatic 'kidnap' during which they may be subjected to cold analysis or physical examination by non-human entities. Ms Tessman claims she was abducted from a farm in Iowa at the age of 7 and told she was special, but that certain aspects of her life and her mission would be hidden until the time was right. These are very common themes.

In her book *The Transformation*, Diane Tessman explains that she was eventually encouraged by Tibus to establish a group and a magazine to publicize his channelled words to those on earth who were able to respond. There are reportedly thousands of 'starseeds' who instantly resonate with these warnings about climatic and physical changes through which the earth must pass.

American author Brad Steiger believes that there are 'star people' scattered throughout the planet: possibly millions who do not realize consciously they have had alien conditioning, or even have alien heritage themselves. This idea is filtering through to other aspects of the supernatural, where researchers using hynosis to find past life memories are now beginning to uncover cases where people recall existences as an alien on another world. The work of psychologist Dr Edith Fiore is an example. There is rarely – if ever – an overlap, but the names of these beings and their planets have the same science-fiction comic-book ring to them, suggesting that their origin may be nearer home than outer space.

Brad Steiger, a popular figure in the New Age movement, and author of many successful books about alien contact, inner-space philosophy and 'Star People' here on earth. (*Jenny Randles*)

One school of channelling thought says that the aliens not only live on other planets but are also from our distant future, taking part in some joint experiment in mind contact. This has an almost plausible air to it, although it still does not overcome the basic question often posed by sceptics: why don't these channellers offer something undeniably unknown to our present technology, for example, a cure for AIDS. If they are receiving messages from advanced beings, this ought to be quite simple and would end the research and debate overnight. Instead, the messages they most often convey are precisely akin to the pop mysticism gurus throughout the ages have spoken.

Jenny met quite a few channellers in Arizona and concluded that some were unaware that they themselves could be the source of these messages, rather than merely the medium through which the information passes. Of course, those she discussed this with rejected such a suggestion with sympathetic restraint. It may be that they do have a channel to another world, although from the glibness and practical uselessness of almost everything the channellers have to offer, it is difficult to believe this is a very advanced civilization.

In one channelling session Jenny contemplated attending, payment of the enrolment fee, which was a fairly substantial number of dollars, would have bought the dubious compensation of a certificate with details of the planet on which Jenny had been born. Wicked thoughts of paying out hard-earned cash for a completely genuine piece of paper which might have simply borne the one word 'Earth' were sufficient to put an end to that consideration! But other people went and doubtless got satisfaction for their money. Who can prove that what they received was not also some form of truth – albeit subjective truth?

One of the biggest boosts to the channeller movement in recent years has been the support from the Oscar-winning actress Shirley MacLaine. This was first presented to the world in her delightful autobiography *Out on a Limb*, and then in the seemingly endless stream of sequels. Unquestionably well written and recently turned into a TV mini-series, *Out on a Limb* relates the spiritual quest which Shirley MacLaine undertook and how it led her into the heart of the New Age movement.

Journalist and UFO researcher Antonio Huneeus told Jenny of his meeting with the actress in New York where she discussed her experiences. He also wrote of this in a review of her work for *UFO Universe* (September 1988). They first met in August 1987 at a seminar for the cultivation of the 'higher self' which Shirley MacLaine organized. She proved remarkably knowledgeable about many things, from ancient civilizations and the intervention of aliens in our history, to channelling and modern UFO sightings.

Indeed she even spent a week in Switzerland, staying with farmer Billy Meier, who has taken dozens of controversial daylight photographs and claims to have met a female alien called Semjase from the Pleiades star cluster. In her writings about her trek to Peru to seek enlightenment, another female Pleiadean (reportedly named Mayan) also crops up in Shirley's story. Ms MacLaine also says that she has had experiences herself: such as seeing a UFO in a visionary state while in Peru making the TV movie of her book.

Undoubtedly sincere as Shirley MacLaine is, her warm-hearted promotion of New Age ideals has certainly helped to establish the movement, often to the distaste of serious UFO researchers. They view trance-channelling and visits from Pleiadean spacewomen as highly detrimental to their efforts to force science to take the subject more seriously.

Earlier we mentioned Dr Leo Sprinkle as assisting channeller Diane Tessman. Given his impeccable background and qualifications, his opinions on this phenomenon were well-worth canvassing. He was kind enough to offer much food for thought about his work, which includes annual seminars in the Rocky Mountains. Here contactees, abductees and channellers regularly converge in the hope that this harmonic gathering might produce something positive.

Dr Sprinkle says that it is important that we focus 'not only on the physical and biological aspects of UFO phenomena, but also on the psychosocial and spiritual aspects.' He even thinks that we will eventually have to turn our attention back to those maligned contactee messages and 'argue about the meaning and significance of those. More fun and games!'

In a paper published in *Psychotherapy in Private Practice* (volume 6, 1988) he discusses his views. He reports that he has 'assisted more than 175 persons [updated by him to '200+' in a handwritten addition dated February 1990] who have explored their UFO memories in hypnosis sessions.' Indeed Sprinkle was the man to whom the ultra-sceptical, government-funded University of Colorado project, known as the Condon Report, turned as a consultant on close encounters during 1967 and 1968 when they produced their infamous dismissal of the UFO evidence. Sprinkle's work on a police officer's encounter in December 1967 was one of the very first such experiments ever conducted.

In addition, he has worked with colleagues 'to explore my own memories (dreams? fantasies?) of some childhood experiences' and has concluded that 'I experienced childhood encounters with a space being on board a space craft.'

He adds that after 31 years of research, he believes (but cannot prove) that these craft are real and that 'perhaps we are entering a new phase of UFO research.' He has devised what he calls the PACTS model to account for the abductee/contactee/channeller experience, with the letters each representing a different phase in such a person's life.

'P' is preparation. This involves a family tradition and childhood track-record of paranormal phenomena, with vivid or lucid dreaming often incorporated. 'A' stands for the abduction, where repressed memories, often released through hypnotic regression, hide a kidnap scenario aboard a craft during childhood and early adulthood. This leads to 'C', which stands for contact, where the experience broadens out during adult life into communication with space beings of the sort that channellers frequently describe, but which he terms 'psychic'. The following stage – 'T' for training – may not always be fulfilled. This is why Sprinkle sees the need for the involvement of counselling. The person starts to obtain information on UFOs and other phenomena and may change personality. This often reflects the first awareness of an implanted message that is waiting to be triggered within the subconscious. The ultimate phase is 'S' for service. Here the person may simply help humanity by spreading the word, or they may develop their inherent or acquired psychic skills in such ways as healing the sick. Nowadays a growing trend is in the semi-preacher mode of channelling, encouraging spiritual messages from these beings to prepare the world for the problems that it must face.

This vision of the phenomenon is an extraordinary one. Perhaps Leo Sprinkle has found a unifying principle that will link these diverse and meaningless messages from UFO contactees to space-message channellers into one remarkable scenario. If he has, only time will tell.

One can understand why the hopelessness of our predicament, or the inability of science and government to rid the world of evils (indeed at times to be primarily responsible for such evils), may have prompted the idea that we need to look beyond the mundane for rescue. The danger is that if we sit and wait too long for the intergalactic cavalry to arrive, when the bugles sound there will be nobody here and we will have sacrificed our future for little more than a pipe-dream.

For the sake of our descendants, we cannot afford to take the gamble that there is a real 'space brotherhood' waiting to ride to the rescue. If there is, then one would hope they might expect us to try to help ourselves and not sit by and expect them to bail us out. If there is not, then at least we will have done more than gazed hopelessly into space, swapping incomprehensible and trivial cosmic messages among ourselves while the planet simply fell apart around us.

11
Inside the Aetherians

Most people are unaware of it, but apparently Jesus Christ momentarily returned to earth on 23 July 1958. He arrived in Britain from Venus by flying saucer and landed on Holdstone Down, north Devon.

The reported purpose of his visit was to meet Londoner George King and hand over a fresh series of teachings, intended to supplement the Sermon on the Mount. These 'Twelve Blessings' have since regularly been recited by thousands of people in Britain, the United States, Canada, Australia, New Zealand, Nigeria and Ghana – members of the organization King created: the Aetherius Society.

Among the channellers and contactees who number thousands around the world, the Aetherians are the most remarkable for their longevity – despite changing fashions. Many people in Britain were introduced to the society in the late 1970s, during a TV documentary on UFOs made by the BBC.

The programme was filmed on location – the wind-swept hill of King's alleged encounter with Christ. A weird contraption strapped to a surveyor's tripod was the centre-piece of the opening scenes. Looking like something from the props department of the TV science-fiction series *Dr Who*, this was a 'prayer battery', a device capable of storing 700 hours of spiritual energy for up to 10,000 years, it was claimed. Its design, which includes elements such as gold, was worked out by George King himself, in conjunction with various 'space intelligences'.

Then appeared the Aetherians – a long line of men and women dressed in jeans and anoraks, faces devoid of expression, many wearing sunglasses. Muttering a deep rhythmic mantra, they approached the tripod one by one, '*Om mani padme hum . . . Om mani padme hum . . .*'.

Just how did it all begin? The story is revealed in a self-published 1961 book, *You Are Responsible*! The tone is set on the cover with a quote from 'A Man From Mars'.

It tells how in 1954, a 35-year-old London taxi driver, called George King, had a strange experience in his Maida Vale bed-sit. From nowhere, a voice commanded: 'Prepare yourself. You are to become the voice of the interplanetary parliament.'

He was already practising yoga, successfully attracting the attention of an Indian sage who teleported around the world to visit his London flat.

Richard Lawrence, leading spokesperson for the Aetherius Society in Europe, discusses the esoteric side of alien contact before an audience of millions on BBC television.
(*Jenny Randles*)

Accepting the position, King became a direct-voice medium. This meant he allowed space aliens and even Jesus Christ to take over his body and vocalize using his larynx. These contacts have continued for 38 years, and many can be purchased on cassette. However, King's first contact was with an alien from Venus given the pseudonym of 'Aetherius', hence the name of the society.

These days, George King, or 'Sir' George as he is known, has escaped from the draughty hills of Devon to more sunnier climes in California, where his organization flourishes. However, his European Secretary, Dr Richard Lawrence, was available for interview by Peter Hough at the society's headquarters in south London. Lawrence is a polite, informed and very lucid speaker, who talks enthusiastically about the 'prayer battery'.

'With this, we can do something which the Vatican cannot. Two hundred members, highly trained in Buddhist mantra and Christian prayer, regularly meet in London, Los Angeles, Detroit, Auckland and Barnsley, Yorkshire, charging up the battery with great prayer energy. In a matter of minutes, this beneficial energy can be directed to any location in the world. We responded during the Mexican earthquake, and many people who should have died were brought out alive from the rubble. During my successful Australian tour in 1978, one commentator remarked I had received more publicity than Elizabeth Taylor the week before.'

How had Richard Lawrence become involved with this strange religious group based on alleged contact with extraterrestrials?

'I became associated with the society at Hull University while studying for my bachelor of arts degree. Already a member of the Buddhist and Vedanta Society, I suppose I was searching for the truth. After attending a lecture given by the Aetherius Society at the university, I decided to join.'

Apart from being psychic, Lawrence also claims contact with 'astral entities'. But Peter was more interested in where his doctorate had come from. He indicated a plaque on one wall.

'It's an honorary doctorate of divinity, awarded for my work in Australia and New Zealand by the International Theological Seminary, based in California. A legally registered evangelical institute. I recently sat and passed an examination for a Doctorate of Theology from the International Theological Seminary in Van Nuys, California. Sir George has one too.'

This was not surprising, as George King has acquired an amazing catalogue of grandiose sounding titles over the years. Apart from his 'doctorate' and 'knighthood' (conferred by the Byzantine Order of Saint John, not the British Queen), he is also a 'reverend', 'archbishop' and is sometimes referred to as 'His Eminence Prince George, King de Santorini, Count of Florina'. He also has a doctor of science degree in 'Astro Metaphysics and the Cosmic Sciences from the North West London University'.

'The North West London University' was never a recognized part of the standard British educational system. Neither *The Compendium of University Entrance Requirements*, nor the *UCCA Handbook*, list it. In fact, it was a private university which closed down in 1988, after legislation passed by parliament outlawed it.

Lawrence believes that 'gutter Press' have given the society a hard time with their suggestions that George King only set up his organization to serve as a self-financing, money-making racket.

'I'm not going to drag out examples, I'd rather they died. If they think we're in it for the money, perhaps they're judging us by their own motivations. In fact Sir George is a prime example of dedication and has never taken a penny in royalties from all the books he has written. He lives in a very modest way.'

Where does he obtain money from then?.

'He draws a salary, but he doesn't take royalties, that's what I'm trying to tell you! But he lives modestly, compared to the way he *could* live. You have a bungalow. So does he – in Santa Barbara.'

Peter smiled inwardly at the comparison between his semi-detached house in down town Warrington, and King's bungalow set in the foothills of California. Richard Lawrence leaned forwards, and continued earnestly.

'I realize you must investigate the possibility that the claims of the Aetherius Society are not true, that either Sir George King is deluded or he is making it up. We don't expect people to believe us just because we say it's true. There are frauds around, and people who fall victim to lies. But Sir George is capable of bringing outstanding revelations of great truth.'

A reference to King's humble beginnings as a London cabbie elicited this response:

'Are you trying to say there's something wrong with taxi drivers? Do you find them fraudulent? Jesus was a carpenter. Does this lessen Christianity? I'd rather listen to the carpenter, Jesus, than the philosophy of any High Court judge.'

In *You Are Responsible*, King describes his first visit to Mars in an out-of-body state, where he falls foul of a deadly dwarf with a ray-gun. Fortunately for the future of the Aetherius Society, he survived to tell the tale. Then, like a hero from *Star Wars*, he was commandeered to help the Martians destroy an intelligent meteorite which was wiping out their space fleet, and ultimately threatened the earth. Even the Martian kamikaze pilots failed to eliminate this sentient lump of rock 'the size of the British Isles'. King led a final make or break attack and defeated it 'with a weapon of love'.

Did Lawrence believe all this was credible? He smiled at me pityingly.

'Obviously I do, or I wouldn't be here, would I? If I didn't believe every book that we publish, and every word Sir George has uttered, I would resign. There are priests in some other religions who don't believe in the resurrection. I think that's dishonourable. I believe everything about the Aetherius Society and Sir George King.

'The fact that some people poke fun at us is very childish considering there is a serious issue at stake. When you talk about Martians communicating, you come right up against prejudice. Most people don't understand that the aliens operate on many frequencies; they are able to manipulate matter, and occasionally operate on our physical plane. That's why all this stuff about nothing being able to live on Mars and Venus is a nonsense. Spiritual entities don't need to breathe oxygen.'

Surely, this is the crux of the argument? On the one hand the society is saying that the aliens are spiritual, and on the other it talks about them 'manning' spacecraft and using nuts-and-bolts weapons to fight off other, evil extraterrestrials. If they were spiritual, what need would they have for physical objects? Indeed, how would they construct such things?

'We believe that life exists on many levels and frequencies of energy manifestation. This is why some UFOs are reported as disappearing and then reappearing again. On each of these levels there are 'nuts and bolts' apparatus, including spacecraft and weaponry. They need such craft in order to travel for certain specific space probes, and to conduct scientific research. They have control over matter to the point where they can travel through the frequencies, including our own. They construct physical objects on whatever plane they choose through advanced scientific methods, including thought control'.

The nuclear debate is currently more active than at any other time. With the release of government documents in 1987 about the Windscale accident 30 years before, the Aetherius Society was not slow in making the most of it. It had been announced in its journal, *Cosmic Voice*, two months after the incident in October 1957, that 'Mars Sector 8' had informed George King of a huge cover-up by the authorities; that the escape of plutonium dust, radioactive iodine and strontium 90 had more far-reaching effects than was being admitted. This has turned out to be true, although many others, who were not members of the society, thought much the same at the time.

The society also cites the case of exiled Russian scientist Dr Zhores Medvedev, who revealed in *New Scientist* in November 1976 that there had been a hushed-up Soviet atomic accident in 1958. The society has pointed out that it broke news of the accident in the June/July issue of its magazine *at the time it had happened*, 'through information given by Mars Sector 6 and the Master Aetherius'. *New Scientist*, rather tongue in cheek, later admitted it had 'been scooped by a UFO'.

Had it been? The evidence is pretty convincing, yet can we be sure that some inkling of these events did not leak out of the USSR through more conventional means?

Despite its rigorous campaign for acceptance, the Aetherius Society has also received a good measure of antagonism from objective ufologists. In recent years the Aetherius Society has pursued a vigorous campaign in the media to encourage the public to report UFOs directly to it, via a special hot-line number. Why, if the Aetherians already 'know' what UFOs are, do they collate fresh reports?

'If someone experiences something for which they have no ready answer, who can they report it to? If they contact the police, the police are prevented from commenting. If they report it to the MOD, they say no defence implications are involved, so they're not interested. There's no education programme in our universities and no proper organiza- tion. We are the only organization able to offer any kind of intelligent discussion or advice on the subject. That's why we do it – to help the people.'

Of course, it is nonsense that 'there's no proper organisation' to which UFO experiences can be reported. There are plenty of organizations which can handle calls from the public and offer sensible advice not coloured by religious beliefs. The British UFO Research Association has a hot-line number in London, and there are others, such as the Independent UFO Network and the Manchester UFO Research Association. These organizations will pursue such reports objectively. The Aetherius Society begins with the premise that UFOs are spaceships piloted by ethereal beings from Mars.

But the Aetherius Society claims to have photographic evidence of their space contacts. Stuart Henderson of the society took a picture of the sunset over Warrington in Cheshire. When a print was made, it also showed an unusual image in the top right-hand corner. The late Mr Henderson claimed the shot was taken through an *open* window.

However, we have examined several photographs in the past which show similar effects. In those cases the image has been caused by the reflection of a house light on glass. Indeed, where the image is less intense in the Warrington picture, background cloud cover can be seen through it. We have actually talked to Mrs Henderson and she told us of the many golden UFOs which have materialized over the Manchester Ship Canal close to their home. We have investigated other reports over Warrington of similar phenomena. In these cases, witnesses were just observing aircraft coming in to land at Manchester Airport.

Perhaps the answer is more simple than contact by Martians and Venusians. The Aetherius Society obviously fills some kind of emotional niche in certain people's lives. This crutch of belief that there are nice friendly space aliens who are willing to help us out of our own difficulties is obviously a comforting thought. As Professor J. A. Jackson, of Queen's University, Belfast, said in a long essay when commenting on the emergence of the group: 'Neither the teachings of the Churches nor the wisdom of science could give authoritative answers. The Society offers certain knowledge and a direct means of contact with space people themselves.'

Yet the 'space people' believed in by the Aetherius Society are more likely a product of the *inner* space of the mind than outer space where there might lie genuine extraterrestrials.

The Search

12

Somewhere over the Interstellar Rainbow

In 1985, Glasgow University astronomer Professor Archie Roy was in buoyant mood. He told a journalist from the London *Observer* that, with new efforts to search the universe for intelligent signals, 'we can expect to make contact very quickly, probably within a decade.' He added that he thought civilizations were 'ten a penny' in the cosmos.

A year later, in an interview with Paul Whitehead in *Flying Saucer Review* (volume 31, number 3, 1986) Professor Roy confirmed this view by saying, 'if we are the product of natural evolution, it is highly improbable that we are alone in the universe.' Presumably this leaves the door open just in case we are not solely the product of natural processes (as scientists understandably assume), but are also the creation of a mystic force, otherwise known as God.

Roy actively pursues his broadly based interest in this search. He subsequently became associated with *Flying Saucer Review*, and he has also become an active researcher and spokesperson in the heated debate over the potential 'alien' messages said by some to lie behind those crop circles recently found dotting the rural landscapes of our world.

However, the astronomer's seemingly reasonable hopes are, as yet, a long way from being fulfilled. Contact is proving unexpectedly elusive, which has led to some quite contradictory statements.

For instance, in 1981 Michael Papagiannis, of the astronomy department at Boston University, said that:

The euphoric optimism of the 'sixties and early 'seventies that communication with extraterrestrial civilizations seemed quite possible is being slowly replaced in the last couple of years by a pessimistic acceptance that we might be the only technological civilization in the entire galaxy.
(Royal Astronomical Society journal, volume 19, pp.277–281)

One can hardly find more polarized opinions than these, and they represent a crucial debate that increasingly dominates the field. While there seems to be a gut reaction based on deductive logic shared by most scientists, implying that life should be 'out there' in great abundance, there is mounting concern at our continued failure to find it.

Long before we understood the universe in any detail, we dreamt about this quest for alien life, and, as we have seen, still speculate on what forms such beings might take. When science fiction became popular during the last century, we even began to wonder how we might establish contact.

Early ideas were ingenious, but impractical: such as building a giant mirror and using sunlight to send Morse-code signals to the (then still plausible) inhabitants of the moon or Mars. Of course, the limitations of physics meant that this could never work, even if there were Martians to see the signals. Only the brightest light that we can produce (a nuclear explosion) is potentially visible from another world and this lasts such a brief time that it is hardly likely to produce incontrovertible proof of life on earth. Alien scientists would dismiss any sightings just as freely as ours now reject claims about UFO appearances.

Another problem concerned the code to be used. How could the Martians have recognized the message, even if they had been able to see it? To them it would have been a meaningless series of flashes. How would they have unravelled any meaning behind it?

This problem exists even if it is assumed (as it nearly always was back then) that Martians, although probably looking like bug-eyed monsters, would still think like human beings. The truth is surely that aliens would be alien in every way and their thought processes would not work in the same manner as ours. That said, the chances of any message from us to them being remotely comprehensible appear to be feeble.

In science-fiction stories and films, such a problem is largely ignored, but that is merely an expediency to help the plot along. We suspend scientific logic to accommodate the story line. However, in any real search for life in the universe, we cannot afford to ignore such scientific reasoning. This complicates matters so much that one or two researchers even think it is a forlorn task. We will never communicate with an alien intelligence, even if we do come across one by chance. The result will be like a farmer staring at a cow and attempting to convey, by spoken language or gesture, why it has to go peacefully to the slaughterhouse.

These problems receive too little attention, even today. Our ability to humanize the aliens is an extreme failure on our part, which academics refer to as 'anthropomorphism'.

Some scientists attempt to get away from anthropomorphism by suggesting that for life to be intelligent it has to be substantially akin to ourselves, thanks to the natural laws of evolution and survival. They also point out an interesting feature of these processes. The Australasian continent was isolated from the rest of the world throughout much of geological history to such an extent that its fauna developed as if it were on a different planet. Yet – the argument goes – there are several parallel life forms: creatures that are totally different in genesis and yet look remarkably alike, despite having grown in total isolation. They are similar because their requirements for survival are the same.

Unfortunately, this logic cuts two ways, for there are many forms of life in Australasia that could have thrived in other parts of the world, but they are unique to that continent. The kangaroo and wallaby are just two of countless examples and, indeed, a colony of wallabies now lives happily in the Derbyshire Peak District of England after a couple were accidentally introduced there a century ago. They have shrugged off all the pressures of modern civilization to become a part of the local ecology.

Furthermore, those few cases of parallel evolution – or similar creatures developing in Australasia and elsewhere on earth – remain the product of a planet with more or less uniform conditions – if one takes into account changes in climate around the world over millions of years, then variations even out. It is unrealistic to extend this reasoning to other planets in the universe, where the environment and evolutionary trends are almost certain to have been very different from earth's.

All evidence suggests that intelligent aliens – if they exist – ought to be so strange we may not even recognize them as being intelligent, let alone be capable of holding a meaningful dialogue with them.

We need only consider the case of the dolphin – a mammal like ourselves with a ratio of brain size to body size so similar that in some circumstances it may even be said to be better endowed. There is a lot of evidence for the dolphin being a highly intelligent species. They have a system of language/communication and a capacity for emotion and even heroic deeds that we presume arrogantly belong only to humankind.

By all reasonable criteria, the dolphin is an intelligence of a similar level to our own. Yet it developed in an 'alien' environment – the 'other world' of our oceans. It is probably the closest we come to firm evidence for what to expect from aliens on another planet. Which is not to say they will look like dolphins, only that they will probably be as different from us as these remarkable co-inhabitants of our earth.

According to Captain David Holmes of the US Navy, here is a description of how a group of dolphins map new territory. Compare it with a human version of the same situation. Scout dolphins go to explore the new region. They then return and convey information via lengthy 'conversations'. A group of 'leader dolphins', who act as a decision-making body, then get together and 'converse', determining whether to avoid, explore, live in or pass through the new area. The rest of the dolphins seem to comply with their advice. Is this not intelligent, social behaviour?

Even more remarkable is Dr Carl Sagan's description of a visit to the long-term, dolphin-communication experiment set up by Dr John Lilly in

1963. Sagan was introduced to one dolphin, which swam up to him and presented its belly to be rubbed. Sagan obliged. The dolphin swam away and returned just beneath the water, so that Sagan had to dip his hand under the surface to provide the pleasing sensation. This occurred again, with the dolphin being even deeper under water. Finally it returned so deep that Sagan would have had to dive in to make contact and he declined the invitation. The dolphin, which had been taught to utter a few passable imitations of human words, showed its feelings by leaping high out of the water and facing the scientist, shrieked one of these imitated words – 'More!'

Anyone who has been close to dolphins in the wild or captivity knows that they are gentle, quite wonderful beings with distinct personalities. We understress their intelligence purely because it manifests in ways quite different from our own – also because of our own desire to feel omnipotent. Dolphin society has no cities or technology; equally it has no wars, pollution or starvation. It is very much a moot point as to which civilization might be regarded as the most advanced on earth by any visiting alien race!

Perhaps we should establish a two-way contact with the dolphin before we start to talk to anyone from another planet. If we do, it will be equally historic, because the dolphins are, quite literally, alien intelligences living right here under our noses.

It is interesting that Carl Sagan has become one of the world's best-known authorities on cosmology after being the originator and presenter of the acclaimed TV series *Cosmos*. John Lilly is probably best known for his strange experiments into the nature of consciousness. His book *The Centre of the Cyclone* describes some quite peculiar communications with 'other intelligences' during altered states. He was a pioneer of the quest for other life within 'inner' space as opposed to searching outside in the universe. His evidence is in many respects the forerunner of today's UFO abductee and channeller material.

Yet almost 30 years ago these two men, now at opposite extremes of the trail for contact with other intelligences, came together for one simple experience in a dolphin pool which may say more about the problem than anything else they have discovered.

In focusing outwards in our search for other life, we may, or may not, be making a fundamental error. However, it is still what most scientists choose to do, so we must examine the justification for their work.

To decide whether other life forms are 'out there somewhere', ready to be contacted, we need to answer four basic questions: How many stars are there? How many planets do they possess? Will there be life on any of these worlds? Will it be the sort that might want to establish contact with us? We are gradually learning more about each of these problems.

The first one is the easiest to dispense with. It has long been known that the universe consists of, quite literally, countless billions of stars – one estimate is a minimum of 10 followed by 21 zeros!. The number is so huge that it becomes statistically meaningless to write it down. Analogies, such as 'more stars than there are grains of sand on the beaches of the world', begin to convey some impression of the immensity.

When we look at the sky in a smog-free environment late at night, we can see no more than a negligible fraction of the full total. Some are quite close (in cosmic terms), others inconceivable distances away – requiring tens of thousands of years even for light to traverse them. Indeed, stellar distances are measured in what are called 'light-years' – one light-year being the billions of miles light travels in 12 months. Even the nearest star (part of the Alpha Centauri formation) is 4.2 light years away – or well over 50 *million* times further than the moon.

When you bear in mind that it took us decades to plan one trip to that uninhabited rock and a week for three astronauts to get there and back, even at what are quite fantastic speeds by comparison with Concorde or a supersonic jet fighter, then it is little wonder some scientists question whether interstellar travel will ever become possible.

Of course, we will develop new methods of propulsion and with as yet undreamt-of technology may one day achieve significant fractions of the speed of light. However, we will probably never get close to that speed itself (which current physics says can never be equalled). Unless we are very, very wrong in our appreciation of science (and this seems most unlikely) then the sort of 'warp speeds' one meets in *Star Trek* will never occur. Even thousands of years into our future we will be forced to crawl to a star as close as Proxima Centauri at such a 'slow' rate that it will still require far more than a human lifetime to get there and back.

Faced with such significant (probably insurmountable) barriers, it may be that actual communication between intelligent species is quite simply impossible: no amount of advanced technology will ever make it happen.

Of course, we can dream up ways around this obstacle. However, the basic question is answered: there are many stars out there and each one is a sun like our own – but they are incredibly, unreachably distant.

All stars share certain characteristics. They are known to be gigantic nuclear furnaces which burn their fuel to create heat and light for some tens of thousands of millions of years, eventually dying in one of several ways. According to various factors, such as size, power, age (called 'position on the main sequence' – a kind of life graph of stellar evolution), they range from very hot giants to almost cold dwarfs.

Our sun happens to be a middle-of-the-road star, which means it is of average size and intensity. It is no more than a very ordinary, quite humble speck within a universe filled with many grander sights.

Not all stars could even conceivably host intelligent life. Those that are too cold or too hot or are unstable would almost certainly have never had the right conditions for a sufficiently long period to allow life to develop. So middle-order stars closely similar to our own sun are the ones that we must assume to be the most hopeful candidates for alien life. Fortunately, there are still millions upon millions of these; enough for statistical odds on their own to suggest that life must be certain.

However, there are other factors to be taken into account, which brings us to our second question – how many suitable stars have planets? It is imperative to consider this, because even if there were a star out there absolutely identical in every way to our own sun, without planets (i.e. a solar system), there would be nowhere for aliens to evolve.

Until quite recently there was no way of obtaining direct evidence of such a thing. Even if the nearest star, at just 4.2 light-years, possesses a solar system like our own, it would be totally invisible to the biggest telescope in existence. The distances are so great and the planets so relatively small that they are impossible to see. However, there is a more significant problem. The light from the star itself is so enormous that it would completely swallow up any light from planets circling around it.

The Hubble telescope (put into orbit by the Space Shuttle in April 1990) may help in this quest. But even here the chances of seeing a giant planet around a very close, fairly dim star remain slim. Bigger telescopes in orbit or on the moon (where our atmosphere cannot limit their optical efficiency) may at some future date resolve actual images of big planets that surround nearby stars. However, that is still into the future.

Consequently, for most of the time we are forced to rely upon theory to speculate whether there are any other planets out there around the myriad other suns. After all, if our solar system was an astronomical fluke, then all arguments become academic. In simple terms, we would have to say: no planets round other stars, then no intelligent life will exist but for ourselves.

A debate raged between two types of explanation for the creation of the solar system. One said that it was a freak event, the other that it was the normal way of things in the universe.

The theory of the freak result suggested that two suns or a sun and a comet nearly collided in the very early days of the universe. Proto-matter was pulled out from one sun by gravity which eventually formed, by collapse and cooling, into the family of planets we call the solar system. This event would be so rare that it would make other life very improbable.

However, the theory which became more accepted as we explored space was that all stars had a crust of matter which accumulated under the influence of gravity and rotated as they rotated. In those first millions of years after the galaxies formed, this gradually cooled to become the planets revolving around the suns, rather like children attached to a maypole.

It was hoped that reaching the moon would resolve the origin of the solar system, since rocks could be brought back from that distant age. Unfortunately this proved over-optimistic, although the rocks gave us evidence broadly in line with the view that solar systems are common when new stars form.

Luckily, further clues are also available to support the theory. For instance, in 1931 Otto Struve and Christian Elvey noted that the middle-order stars (of which the sun is typical) rotate much slower than the hot ones at the end of the main sequence. We knew that the rotation of our sun was slowed because of the debris trailed in its wake – that is, the constituents of our solar system. This is rather like an ice-skater, arms tucked in, spinning round at great speed, whereas the minute she puts her arms outstretched, she slows right down. The physical principle, known as conservation of angular momentum, causes this just as it slows down a shot-putter who spins round with a heavy weight on the end of a cord. In stellar terms, Struve and Elvey's discovery was important evidence, even if indirect, that most middle-order stars probably rotate slower than expected because they also have a family of planets in attendance.

More information came from observing the motion of stars relative to one another. Because some stars are quite near and others very far away, the closer ones can move more noticeably than the distant ones if you observe them across long periods of time. The same principle applies if you sit on the beach watching a ship far out to sea and some holidaymakers on a pedal-boat close to the shore. Both may be moving, but over a few minutes of time the distant ship will hardly seem to budge, while the pedal-boat will cover quite a wide arc of the horizon.

When watching stellar movements with the better telescopes available in the mid-nineteenth century, some strange anomalies were noted. You could use mathematics to work out the motion that a star should display against the background, but some of them were not behaving as predicted. They 'wobbled' distinctly in their path. The only real explanation for this was that there was something missing from our knowledge about the star system in question, such as another massive object whose gravity had to be added to the equation because it was visibly distorting the motion of the parent star.

Wobbles were first discovered in the star 61 Cygni (1838) and Sirius and Procyon (1844). As the years went by and observations became sophisticated, more and more stars were found to have wobbles.

The cause of these wobbles often became obvious when astronomers were finally able to see a small companion star close to the major one. The second star was creating the distortion. This was the case with Sirius. Double stars and even triple systems, such as our near neighbour Alpha Centauri, are not uncommon in the universe, so this was no surprise.

However, in some instances the reason for the wobble was not so obvious. Even when we had equipment capable of showing very dim companion stars, no such star was found. In these cases something else had to be responsible for the distortion in the path of the star. It was concluded that this extra mass was too dim to be seen if the sunlight from the star swamped it. Then the answer became self-evident to some astronomers. This missing unseen companion was a planet, possibly several. With the help of mathematics, it was concluded that there were other solar systems around some stars relatively near our own. This was a highly significant deduction.

Most of the stars now identified as having solar systems are very poor candidates for life. This is because, compared with our sun, they are small and dim (known as red dwarfs), with limited energy output to warm any planets in their vicinity and so generate life. This small energy emission is one reason why we can be so certain that their wobble is caused by planets and not by small companion stars, which would – if they existed – emit sufficient extra energy to be detected as stars against their feeble neighbour. We can usually find these companions in such dim surroundings but could not be so certain where the intense light from the primary sun blots everything out.

The best illustration of this problem is Barnard's Star. At six light-years it is one of the very nearest stars, but it is so dim that it was not discovered until 1916. Even with the naked eye we can see brighter stars that are many hundreds of times further away.

Barnard's Star was photographed for years and the wobble that it displayed was finally decoded by astronomer Peter van de Kamp in 1963. At least one big planet (of Jupiter size) was said to circle this star. In fact, in more recent research there is now considered to be two 'giants', one almost as big as Jupiter and the other about half its size rotating around the sun in orbits of 11.7 years and 18.5 years. But even if there are smaller earth-like planets nearer to the sun (which would have no noticeable effect on the motion of the parent star and would be invisible from earth), it is considered extremely unlikely that life exists.

Shortly after the discovery of this first confirmed, distant solar system, Jenny took an astronomy course at Manchester University under the guidance of Professor Zdnek Kopal. She recalls how he described this tremendously exciting breakthrough and his words of caution that it signalled a new era for humankind. The astronomer warned that before long we may well find a planet that, unlike the ones in the Barnard system, was inhabited. An advanced civilization seeing our warlike nature may decide to eliminate us before we got to them. This was just an imaginative scenario, of course, but he spelt it out with chilling words. 'If we hear that space phone ringing,' he told the class, 'then I strongly suggest that the earth does not answer.'

Finding other planets beyond our solar system was quite possibly the most important scientific discovery of the 1960s. At a stroke it meant that the chances of intelligent life elsewhere in the universe rose from a possibility to a near probability. For if there are billions of planets out there, as this discovery implied, the odds are that some of them will be earth-like in form around sun-like stars.

However, we still need to know if these worlds would be inhabited. It is not known exactly how life began on earth, but it is now believed that this was basically the product of a natural chain of events. For most of the history of our planet, this produced little more than microbes, then bacteria, later algae and complex plant life, and eventually very primitive animals. Indeed, if you represent the entire geological history of earth as taking 24 hours, intelligence seems to have arrived in recognizable form at just a few minutes before midnight. Intelligence of the order of human beings constructing language, science, technology, etc. (even of the most basic kind) does not appear until just a few seconds to midnight. In other words, for 23 hours 59 minutes and some seconds, the earth was either dead or populated by very primitive life forms. Suddenly, virtually out of the blue on a cosmic scale, we appeared and developed our societies and effectively took over the world!

This fact bears with much contemplation. However, it may be mitigated by our almost total lack of understanding as to the nature of consciousness. We still have no real idea how our 'mind' differs from that of a more primitive species, or indeed whether other species have any mind at all, or whether everything that lives possesses consciousness; at that level we thus differ only slightly – if at all – from other life forms. However, it has certainly provoked some people to suggest that it constitutes a scientific miracle that has yet to be explained. This is why physics has been unable to make God redundant. It is also why some people contend the best non-supernatural solution is that human life is an experiment in genetic engineering performed on virgin soil by an alien race. We arrived on earth overnight because we were planted here!

We cannot answer these questions, but we do know that the basic building blocks from which the stuff of life is formed are found all over the universe. We also know that primitive organisms can exist even in space itself. The Apollo 12 landing on the moon in 1970 found the *streptococcus mitis* bacteria still alive on the lens of a camera sent there two years earlier by an unmanned probe. This bacteria came from earth, but it had survived in an environment far more hostile than the one on many planets that are likely to exist in other solar systems – without air, with heavy radiation exposure and with temperatures swinging wildly between searingly hot and bitterly cold.

Another fascinating clue to the genesis of life comes from meteorites, which are bits of debris from the solar system (apparently dating back to its formation). They enter the earth's atmosphere, partially burn up through friction but can reach the surface, where they impact. They are gifts from space to add to our understanding of the universe.

Nearly all of these bits of debris are fairly mundane rocks, but about 3 per cent are called Carbonaceous Chondrites because they are very different, containing carbon and silica – the stuff of life. Studies of some of these have revealed quite complex amino acids and other molecules. They are definitely not earthly in origin, and, despite some claims to the contrary, they do not appear to be life-bearing as such.

Researchers have suggested that life can form in space and astronomers like Professor Fred Hoyle and Chandra Wickramasinghe have built quite detailed theories (see *Life Cloud* and subsequent books). In their controversial view, life on earth (and even modern diseases) may be planted here by debris from space-wandering comets. A not dissimilar concept was expressed much earlier in fictional terms in Hoyle's brilliant, science-fiction novel *The Black Cloud*.

This evidence is contentious, but it shows that complex inorganic molecules from which life develops actually do exist in space. These must surely form (or be seeded) not just here, but also in other solar systems where conditions may later become suitable for germination. Life itself seems to arise when these molecules go one stage further and start self-replication, which is a remarkable step to take. Indeed, in terms of chance, it would be just as probable as a monkey sitting at a typewriter and writing *Macbeth* by hitting the keys at random.

To date we have only studied moon rocks and found them sterile. However, this is a special case as their origin is probably tied to the earth's. We have landed on Mars and conducted tests in the soil of a desert region. Beyond that we have no evidence as to whether the inorganic basis for life (if not life itself) is present on any other world.

Mars has a very hostile environment on which humans could not survive without artificial aid. However, there was some hope of finding basic bacteria or at least some living amino acids (those early building blocks of life). We found none, but no firm decision should be taken as there has been only a very limited study of the planet. Some geologists, too, believe there is evidence that conditions were once very different on Mars. This might mean that fossils of ancient life could exist in the Martian rocks, so one day they might be brought back to the earth for study.

In fact, even the simple soil experiments on the red planet produced an unexpected reaction which at first looked as if it indicated that there were living 'bugs'. However, further results did not verify this and the concensus view is that the tests simulated life through complex inorganic processes. This is much like that other finding: non-living amino acids in certain meteorites. Both results could suggest that complex inorganic molecules exist on other worlds from which life is but the next step up.

What we do not know is whether the evolutionary process towards life naturally reaches the crucial stage found on Mars, but then stops in all bar one in a trillion miraculous cases. Could the earth be that statistical miracle? If so, then we are probably almost alone in the universe – the result of intervention by 'God', or by one of the tiny number of other races who might have resulted from such a freak. Maybe they once came here to carry out their own experiments – building life on a sterile world. If they did visit here millennia ago, then humanity would be the resultant monster of some cosmic Dr Frankenstein!

Consequently, while in some respects our recent progress has made the possibility of life elsewhere probable, we still face the hurdle of determining whether life itself is a quirk of fate, a supernatural miracle or the inevitable consequence of certain conditions. Until we answer that question, we will not know whether there are millions of alien civilizations in the universe or millions of potentially earth-like planets that never received the 'vital spark' and are therefore nothing but lifeless ghost worlds.

To a degree, this makes our final question somewhat pointless. But we should bear it in mind nonetheless, because even if life has developed in many other solar systems we cannot assume that it would either want to get in touch with us or be capable of establishing contact.

The species might live in a world perpetually covered by thick clouds, or beneath an ocean. In either case, it would probably have no concept of outer space or therefore develop the urge to communicate with other planets. Any number of other factors – too 'alien' to comprehend – might challenge our belief that other creatures would inevitably show an interest in us. Furthermore, aliens might never have developed the sort of technology we have and probably also suffer the constraints we spoke of earlier which prevent journeys to other stars. In that case, they may exist in blissful complacency, and our reality or non-reality is quite irrelevant to their lives.

After all, it is only during the past century that we have given any real scientific thought to the question of contacting alien life. If you went back in time and discussed the concept with the scientists of the 1700s, they would be mystified or regard it as pointless speculation. We cannot know that any intelligence out there necessarily regards it as vitally important to scour the universe in an effort to find us.

All of these factors greatly reduce the potential number of intelligences with whom we might hope to establish communication. Although complex mathematical equations have been worked out (the most famous by pioneer radio astronomer Frank Drake during a symposium in November 1961), the real truth is that they remain little better than guessing games. Each factor in the equation, has so many uncertainties attached to it that the result − that is, the number of contactable alien intelligences − can best be described as being a number 'somewhere between zero and a lot'!

Being almost certain that there are at least a few other intelligences out there is not much use unless we can devise a way to establish contact. We cannot actually go there, at least for the forseeable future (if indeed we ever can). We cannot use methods such as flashing mirrors. It was only in the 1950s, when the development of radio astronomy accompanied the news that other solar systems now seemed a probability, that a new option became apparent to science.

Dr Frank Drake, before graduation at Cornell University, heard Otto Struve lecture about middle-order stars with slow rotations and what this implied about other solar systems. The subject was in the back of his mind when, in 1958, he joined the radio-astronomy centre at Green Bank, West Virginia.

Radio telescopes, such as the giant British dish then under construction at Jodrell Bank in Cheshire (which for a time was the largest in the world), are not at all like ordinary telescopes. You cannot look through them and 'see' stars. They view the universe in a different way, picking up the energy that galaxies and other bodies emit in frequencies other than that fairly tiny portion which represents the visual range.

They provide a great deal of data and the resulting maps of the sky are often able to tell us much about the universe which cannot be learnt through optical telescopes. The energy being emitted by stars in these non-optical frequency ranges travels at the same speed as light. This means that information picked up now from Proxima Centauri, left there 4.2 years ago, or Barnards Star 6 years ago. Theoretically, a two-way

One radio message received by astronomers was so odd that it made as much sense when transcribed on to musical paper. (See also page 133.)

communication could take many years, even with these nearest stars, but it is still far better than tens of thousands of years needed to send a remote rocket probe there.

Consequently, it was quickly appreciated by Drake and others that any aliens might perhaps use this knowledge to beam signals at us and announce their presence. Rather than send a message, wait for a reply to come and then send an answer back to that, the most likely plan would be to send out a continuous stream of information which contained the answers to questions you may predict that anyone who 'tuned in' would want to ask.

Although the technology to send such messages was in its infancy, it was at the level where we could do it. Some scientists argued that other races may already be sending information our way – all we had to do was search for the signals. In the case of sending information, the process is not just a simple case of pointing a powerful transmitter at a likely star and sending out a message to be decoded. The problem can be likened to a shot-wave radio ham searching for a message he presumed was being transmitted by a person marooned somewhere on an island in one of the oceans. How would the ham find that message?

For a start, he would need to find the frequency being used. There are countless possibilities and many other radio communications to get in the way. So he would probably spend ages twiddling about through different ranges, hoping to stumble on to the call by accident and hoping that it could be heard against the background noise and all the other messages.

Even if this were possible, the ham operator would need to hope that the signal was strong enough to reach him. Furthermore, he would need to know that the message was a message. After all, our islander might not use Morse code. He may have an entirely different code (or language) and a unique concept of what constituted an urgent call. It is by no means certain the ham would understand either of these or be able to spot a message against the background interference or other types of communication flooding the airwaves.

These are just some of the problems, although in truth this example is considerably easier than searching the universe for alien messages. Unlike short-wave communications (which bounce back from a layer in our atmosphere and so can be detected over wide parts of the earth), messages from star to star are more directional. You not only need the right frequency (of which there are far more to choose from than with short-wave band on earth), but you also need to be looking in the right direction.

If our radio ham, in addition to all his other problems, had to know which part of which ocean to scan – to look for a message he may never understand or recognize, on a frequency he will need to find by trial and error – then very likely he would give up in despair! Yet this is effectively the task Drake and his colleagues set themselves.

The first experiment, code-named Project Ozma (after the princess of a strange far-off land in a fairy tale!) began at Green Bank on 8 April 1960. As there was so little power on this early equipment (and many demands for its services from other researchers), the astronomers focused their attention on just two nearby stars, Epsilon Eridani (10.8 light-years out) and Tau Ceti (11.8 light-years from earth). These were considered good candidates as they were similar to our own sun. Green Bank turned into a frequency of 1,420 MHz (on a wavelength of 21 cm), since this is the natural point at which hydrogen atoms emit bursts of energy. It was felt that since hydrogen is the most common and basic element in the universe, any intelligence would know that fact and realize its position served as a very useful bench-mark.

A lot of assumptions went into this first limited experiment, but it seemed to pay off instantly. Within hours on the very first day, a strong signal was detected. Unfortunately, it soon proved to be unknown terrestrial interference from a then-secret naval base conducting anti-radar tests. Once this was eliminated, the Ozma project went on for a total of 150 hours without any success.

There have been refinements since then. Should the telescope scan a mathematical constant such as pi – the relationship between the circumference and diameter of every circle? This would be universal and basic to any society with mathematics or geometry. Directions in which to point have been modified as well, thanks to new knowledge about nearby solar systems. And there has been a clever suggestion to send (and look out for) coded information which offers bits of data totalling a prime number (that is, a number that can only be the product of two other numbers). In this way the two constituents of the prime form obvious grid sides on a diagram and, when plotted here, the bits of information recreate a visual image (for example, a map of our solar system or a representation of what a human being looks like).

This last trick was first tested on colleagues by Frank Drake to see if they could solve the riddle. Some did and decoded the picture, giving hope that aliens may be able to do the same. This kind of information has thus formed the basis of the few messages that we have sent out towards nearby stars announcing our presence. The date has passed when instant replies from these 'calling cards' were due, so we must assume that they went undetected – if indeed there was anybody there to detect them!

Of course, technology has improved since 1960 and our radio telescopes are now far more sophisticated. However, the techniques used to search for messages is basically unchanged. We may be able to sample more stars and more frequencies at once, but we are still left looking for what is less obvious than a needle in a haystack. Indeed, finding a grain of salt in a haystack might have a better chance of success.

Since 1960 there have been periodic attempts to search for signals from various radio telescope sites. In 1976 at Green Bank, as part of a trial for a big multi-telescope project code-named Cyclops (for which funding was being sought), two scientists surveyed some 659 stars on the hydrogen wavelength. A few anomalous results were recorded, but none repeated themselves or proved to be conclusive evidence of a message.

But are the astronomers wasting their time? The emphasis on radio telescopes for contact presumes that our current technology is either universal, or the best method available for such a complex feat, neither of which may be the case.

A civilization that does not go through the radio-communication phase would probably never think of sending or seeking messages in this way. Even from our point of view, in a hundred or two hundred years time such a method may be regarded as quaint and useless. We have seen how hit or miss this approach is and that two-way communication with neighbouring stars would be difficult because of the huge gap between messages and answers.

At this stage in our science we have no method that would be an improvement. The speed of light seems to be an impossible barrier, although it is prudent to realize one thing here: something *does* travel at that speed – light itself, indeed all electromagnetic radiation. Consequently, it is best to say that the restriction applies only to material things and not apparently to insubstantial radiation fields.

In the frame of reference of a beam of light travelling from star to star its journey does not take years, as we perceive it to do from here on earth. Were the beam 'aware' of its journey, then the laws of physics say it would actually experience *no* passage of time whatsoever. Of course, this seems to be a meaningless statement because fields of radiation such as light are not 'conscious'. However, that is a presumption we should not really make in view of our lack of understanding about consciousness. Suppose this does prove to be an unknown form of radiation and so, like all other energy fields, travels at light speed. If so, then it would truly pass from any point in the universe to any other point in zero time – i.e. instantaneously. In other words, mind or consciousness would be both timeless and spaceless – everywhere at once – which interestingly enough is just what most religions (including Christianity) describe God to be.

It is easy to get lost amid speculative mathematics or philosophy and what seems suspiciously like mystical nonsense, so we will avoid that temptation. However, it is wise to remember that today's mystical nonsense may often become tomorrow's accepted science.

We should realize that an advanced intelligence might find methods of communication that are far better than radio – for example, by harnessing consciousness as a force, if we can understand how thought may operate. It is presently supernatural fantasy to speculate about aliens using the mind to bridge the stars, but it is a good exercise in lateral thinking. It may prove to be no more than mystical claptrap, but if not, then such methods of data exchange would be so superior to beaming messages by radio telescope on the hydrogen wavelength that interstellar communication would probably be going on all around us in great profusion.

We may be in the same position as a tribe on a remote Pacific island where the witch-doctor has just discovered that you can talk to someone at the other end of the beach using two discarded tin cans and a bit of string. Having developed this technology to a fine art, the tribe is trying to devise methods whereby pictures of their latest war dance can be sent through this to another people they believe must exist on a nearby island. Will they understand? What language should we use, they might ask?

The tribe struggles on with this useless dream, oblivious to the highly advanced civilizations in other parts of the world. These people are communicating with one another millions of time every second by telephone, electronic satellite data and TV signals and would laugh at the very thought of using two tin cans in any serious attempt at contact.

Our tribe really needs a TV set, an aerial and some electrical power – concepts far beyond them. But if it were to receive this technology, then it could throw away the tin cans and immediately tune in to the incredible barrage of information, transmissions and communications that go on around it. Indeed, these transmissions have been going on – possibly for many years – while it argued about its tin-can toy and debated pointless questions of tin-can technology that had no solution or validity.

Perhaps when we throw away *our* tin cans (the radio telescopes) and discover the universal method to listen to the cosmos, then we too will find that those elusive aliens have been there all the time. Indeed, perhaps they are just waiting for us to stop playing with technology that – to them – is no more than a scientific cul-de-sac.

Then we can all get down to the real business of interstellar communication.

13
Eye on the Sky

Jodrell Bank science centre is situated deep in the British countryside, not far from the town of Macclesfield, Cheshire, but it is operated by Manchester University. Long before the site is reached its two giant telescopes dominate the skyline, incongruously space age by the side of the ancient oaks clustered across the rolling plains.

Jodrell Bank was the personal achievement of physicist Sir Bernard Lovell, who saw his dream become a reality, despite government indifference to the project. It began as a muddy field, in December 1945, containing some surplus radar equipment loaned from Manchester University. Lovell wanted to carry out research into cosmic rays.

The growing science of radio astronomy captured the former assistant lecturer and he became involved in a more ambitious project: to build the world's largest telescope. Construction of the Mark 1, designed by Sir Charles Husband, was started in 1957. However, without a personal donation from Lord Nuffield in 1960, the whole idea might have been scrapped. The Mark 1, renamed the Lovell Telescope in 1987, was later joined by the smaller Mark 2.

The Lovell Telescope has a huge dish, 250 feet (76 metres) in diameter, which has to be seen to be appreciated. This acts as a collector for radio waves emitted by objects throughout the galaxy. These waves are reflected towards a central tower, turned into electrical signals, and amplified for analysis. This amazing piece of engineering is fully steerable on circular railway tracks, and the dish can be altered to any angle.

Sir Bernard Lovell retired in 1980, and control of the centre, with its 120 staff, was handed over to the equally able Professor Sir Francis Graham-Smith. Prior to his present position, Sir Francis spent seven years at the Royal Greenwich Observatory, in charge of optical astronomy. His main job there was to oversee the implementation of a new observatory on Las Palmas in the Canary Islands.

Does Sir Francis think that intelligent life exists in the universe?

'I haven't really an opinion on that. I'm basically an observational astronomer. I like to work on things which I can actually observe, or think I have a chance of observing. Supposing that *you* postulate that there is some form of extraterrestrial intelligence, and ask me if there is any chance of detecting it?'

Originally called the Mark 1 radio telescope, this mammoth piece of astronomical equipment was completed in 1960. It was renamed the Lovell Telescope in 1987, after Jodrell Bank's founder, Sir Bernard Lovell, who retired in 1980. At 250 feet (76 metres) in diameter, it was the largest dish in the world for many years. (*Peter A. Hough*)

Supposing then, that there are planets around some of the other stars, and on a number of these life has evolved which has developed to a high state of technology. What are our chances of discovering this?

'The first thing you would do is look at your hypothesis that there are planets around other stars. There's a pretty good chance that there are! But we would need to know whether planets form frequently or not. Star systems form out of condensing nebulae [gigantic clouds of gas and dust] and seem to sort themselves out into planets and satellites. But supposing rather than a large number of planets, it is more usual for binary systems to form – one star and one planet. The possibilities of life developing elsewhere diminish drastically. That question involves quite an observational programme in its own right – something the recently launched Hubble Space Telescope might help to answer.

'How did our own planetary system form, for instance? If there are other systems, you have to ask whether or not conditions are conducive to life. Then you start looking at our own system and see that although conditions are right on earth, they are not so right on the other planets! It is very much a fluke that life has evolved on earth. Mercury and Venus are too hot; the others are too cold. This is the only planet which has developed an oxygen atmosphere. You may postulate that a star with a dozen planets might yield one with earth-type conditions, but we need to learn a lot more before making that kind of deduction.'

But does life have to be carbon based?

'It is easier to build things based on carbon. The energy levels involved in transition from one complicated chemical compound to another, isomeric changes, and changes of excitation of light – these are all the right levels for carbon. No one really knows how life can be manufactured otherwise. I suspect there really is no other way of doing it.'

Professor Frank Drake started off the Search for Extra-Terrestrial Intelligence (SETI) in America, over 30 years ago, using a small 85-foot (26-metre) telescope to listen out for radio signals from possible extraterrestrial sources. Now the National Aeronautics and Space Administration (NASA) are set to start a systematic search in October 1992 – 'Columbus Day'. The project has been funded to the tune of $100 million dollars, over a 10-year period. Did Sir Francis think this was a viable way of searching for life out there?

'It's by far the best way. Although I think opinions are divided on whether or not you want to spend your life searching for extraterrestrial life. But I quite see its attractions, and there's no doubt that if it produced a positive result, it would be very exciting indeed.'

Does Sir Francis think Jodrell Bank should become part of the SETI programme? NASA recently extended its programme to the southern hemisphere with the addition of a 230-foot (70-metre) dish near

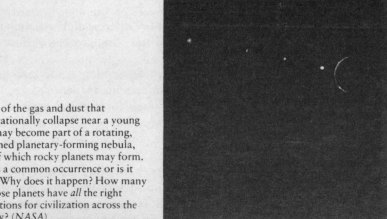

Some of the gas and dust that gravitationally collapse near a young star may become part of a rotating, flattened planetary-forming nebula, out of which rocky planets may form. Is this a common occurrence or is it rare? Why does it happen? How many of those planets have *all* the right conditions for civilization across the galaxy? (*NASA*)

Canberra, Australia. The others are in California and Madrid, Spain.

'No – but for a very good reason: we've already got enough to do. NASA did actually attach a 65,000-channel receiver to a telescope here for a while. We suggested they should try it at Jodrell to see what an intensive area of radio interference is like to work in! It was hooked up to a computer, and lo and behold, it found some interesting signals, although nothing conclusive. The problem was coping with them. I don't think they would want to come back here. They'd try and find a place with less radio and electrical activity.'

Indeed, scientists from the Jet Propulsion Laboratory in Pasadena installed a 65,536-channel pipeline FFT spectrometer in the autumn of 1983 at Jodrell. The test ran for 10 weeks and, according to SETI scientist Dr Jill Tarter, provided a very illuminating look at a situation that will only grow worse. She said in a paper published in 1989: 'Radio frequency interference poses one of the most significant technological challenges to any ground-based SETI programme, and may already have rendered unusable much of the terrestrial microwave window at many sites around the world.' She concluded that future SETI projects might require earth-orbit or lunar-based facilities, which would, of course, escalate the cost.

But Sir Francis has more pressing domestic issues in mind.

'We're not just here to research either, we teach too. Most of the observational work is done as a postgraduate teaching programme. Students go through a piece of research, get a result, write a thesis, then go out and find a job somewhere. Now, they wouldn't get any results on a SETI programme!'

But the SETI programme is largely computerized.

'Yes, but the use of computers does not mean you are less busy – just that you are able to do more things.'

The 'things' Jodrell Bank is able to do more of are mainly research into pulsars and quasars. Pulsars are collapsed stars that give out very strong and regular radio waves.

'Quasars are even more fascinating. Something really strange is happening in certain galaxies. Entire star systems are being sucked into black holes. That is quite exciting enough for me and most of my colleagues. If we learn to understand what happens to all that matter once it has fallen into a black hole, then we are on course to understand how the universe formed.'

The major project that the centre has been involved in, since 1980, is MERLIN – the Multi-Element Radio-Linked Interferometer Network. The bigger the telescope, the better its resolving power. Instead of building a bigger telescope, you can link up a lot of smaller ones over a

NASA SETI scientist Dr Jill Tarter. She went to Jodrell Bank in 1983 and installed a 65536-channel pipeline FFT spectrometer. The equipment was used to search for extraterrestrial radio signals from space. It was being tested at the Nuffield Radio Laboratories for frequency interference before being installed at radio telescopes around the world. (*NASA*)

wide area. The Lovell Telescope is computer linked with six others, including a new 105-foot (32-metre) dish at Cambridge. MERLIN is able to imitate a single telescope 145 miles (233 kilometres) in diameter.

But both MERLIN and SETI will have their work cut out by the time we enter the twenty-first century. Sir Francis explained why.

'Both the Americans and Russians are bringing out different versions of the same global navigational system. This will be available to civilians, such as round-the-world yachtsmen and private pilots, as well as commercial enterprises and branches of the military. By tuning into a satellite, you will be able to discover your exact position to within 98-feet (30-metres)! Unfortunately, the radio frequencies used by this system are the same as those used by SETI and other radio-astronomy researchers. Worse than this, the Americans and Russians are planning to use two *different* wide-frequency bands. It is a profligate waste of frequency channels to build two separate systems. Aside from that, the system is so useful you can hardly expect it to make way for astronomy.'

Not even for the quest for an answer to one of the greatest riddles of all time: are we alone?

14
The Search for Other Worlds

The most profound puzzle of all is the fact that, whatever we may experience mentally, time does not pass, nor does there exist a past, present or future. In its place is an existence of many universes, overlapping at their 'edges', where 'time' as we understand it is an irrelevance.

(*Other Worlds*)

In magical words such as these, Professor Paul Davies has established himself as one of the most admired popular reporters of modern physics. He has shared with millions of ordinary people the awe felt by his fellow scientists as each new day of the twentieth century has brought fresh discoveries about the nature of the cosmos.

Other Worlds is considered as Professor Davies's finest work. It attempts to set in context for the layperson just how fundamental are the changes to our comprehension of the universe that have been brought about by the quantum physics revolution. While not replacing the old reality of space and time championed by Newton and Galileo, it has introduced some amazing insights that effectively turn our perceptions on their head. Solid reality has become a ghost universe filled with phantoms.

For some years Davies has been Professor of Theoretical Physics at the University of Newcastle-upon-Tyne. However, in March 1990 he emigrated to the warmer climes of Australia to take on the Chair of Mathematical Physics at the University of Adelaide. When we spoke to him he was much looking forward to this new challenge.

Because of his ability to express often complex theoretical questions in easy-to-understand terms, Professor Davies's views on the subject of extraterrestrial intelligence are especially welcome.

'I think there is other intelligent life in the universe.' He said confidently. 'If the universe were infinite then, of course, on probability grounds this would be a mathematical certainty. But my view is based on more than a theoretical opinion. By studying the principles of general organization in both living and non-living systems, we find it suggested that this is a fundamental principle of the universe. But we do not yet know if life would be rare or commonplace even if it does exist, as I feel certain that it does.'

In his book *The Cosmic Blueprint* he elaborates more fully on this idea, showing how nature tends to organize into more complex forms.

Never afraid to challenge convention, while firmly rooted in physical logic, this work, like many of Professor Davies's titles in the 1980s, probed into regions once considered scientific taboo. He has even written a book with the title *God and the New Physics* – a topic that may have seemed heretical in the not-too-distant past.

But if life does exist out in the universe, have we ever seen any sign of it from down here on earth?

'There have been a few oddities. It is great fun to read about them and very intriguing to ponder the potential implications. But – while I am a great supporter of SETI – I do not think that any one piece of evidence yet reported stands on its own as definitive. The discovery of pulsars was a case in point, where signals from a naturally occurring source were initially misinterpreted by some people as evidence of life.

'But I believe that any intelligence out there could make it obvious that what it is sending *is* a signal, not just a natural transmission. If what we detect is just a hiss – pure noise – then it would not be an indication of intelligence. But if there is any regularity, we should be suspicious. A colleague of mine has come up with the idea that the frequency to search is that of hydrogen *but divided by* the mathematical constant pi – not just the frequency of hydrogen itself as many scientists propose. This suggested frequency would show some thinking process – some intelligence.

'But at the moment we have nothing to work with; only a few odd echoes and some false alarms. So, while there is no clear evidence, it remains a possibility that we need to search for. In Australia they are about to begin the first southern-hemisphere search. I am excited about that.'

Traditional science has focused on the radio telescope as the tool to seek out signs of intelligence. But it is argued that this presumes other life would consider radio waves the most appropriate means by which to communicate. After all, we have only used this process ourselves for one century and that represents little more than a blink, even in the history of humankind, let alone the billions of years of life on earth or the probable lifetime of any hypothetical alien intelligence.

Given his penchant for thinking in a radical manner, did Professor Davies believe that we were applying the correct logic by scanning the electromagnetic (EM) radiation spectrum?

'It is true that they could have a totally different technology, but it is the job of any intelligence out there to know what sort of technology we – the seekers – would develop first. Having presumably had the experience of using radio communication themselves, then they would be expected to use this to try to communicate.'

This NASA 230-foot (70 metres) dish, located near Canberra, Australia, is one of three in the Deep Space Network. It will be used to provide southern-hemisphere coverage for the Targeted Search of the SETI Microwave Observing Project. The other dishes are located near Barstow, California, and near Madrid, Spain. The weight of the rotating part of the structure is more than 6 million pounds (2.7 million kilograms). (*NASA*)

Of course, this does rather assume both that an intelligence *has* gone through the development of radio communication *and* that it would want to communicate with an intelligence like us at the earliest stage of attempting to probe into the universe. Perhaps they would rather wait and talk only to more advanced civilizations that have passed the phase of radio contact and any other social problems that may accompany such a level of technology. After all, atom bombs and nuclear missiles are a product of the same era in our scientific progress.

Paul Davies answered carefully:

'As a physicist I know that there is EM radiation in the universe, so it is natural that I would expect an extraterrestrial intelligence to manipulate this to try for contact. Any intelligence should know about the fundamentals of physics.'

'But the other question is: would they think it was the right thing to do? Would they spend the money or take the trouble to blast the universe with radio waves. The arguments for extraterrestrial intelligences doing this are rather more weak. But, on the other hand, it is fair to expect an extraterrestrial intelligence to be curious.

'I do agree that this debate is rather anthropocentric. It is based on what we think we would do, which may not be how an alien intelligence would necessarily behave. They may have totally different ideas. Indeed, this is often used as a response to critics by those who do not find the current lack of evidence persuasive that there are no extraterrestrial intelligences to be found. Unfortunately, until we contact a form of life from off the earth we will not know which parts of our lives are unique to us and which parts are universal in the truest sense of the word. But I tend to think that we are not that special.'

The discussion broadened into a more general consideration of this question. Just what would an extraterrestrial intelligence be like, were we ever in a position to communicate with it? In *Other Worlds* Davies had pointed out that life as we know it is carbon based, but that some biologists think it might be possible for life forms to evolve using silicon or other similar elements as the building blocks. However, these were highly speculative and as yet far from proven possibilities.

'I don't think there is anything very special about the human form. It is easy for us to think that to be intelligent – to develop tools or a technology – you must have two arms and two legs, but I would be surprised if any extraterrestrial was too much like us. They could be very different indeed. Having said that, if they were, they could have very different concerns from our own.

'For instance, think of a planet entirely covered by water. What sort of society might result there? Would it have any technology? This is a very tough sort of question to answer. It is hard enough to comprehend the different species on earth which have degrees of intelligence. So imagine species from a totally different environment.

'Nevertheless, I believe that life is channelled by the forces of nature and evolution, so it would not surprise me if life forms somewhere or other in the universe were very similar to ourselves.'

Finally, we touched upon an area of great controversy about which Professor Davies had his own very pertinent comments. Is there, perhaps, some trace of these missing extraterrestrials in the sort of evidence to which few scientists are prepared to give much thought? What about the seemingly escalating tales of UFOs and alien contacts?

This is something with which Paul Davies is surprisingly familiar. From his wonderfully open-minded pursuit of knowledge he has kept a close eye on the matter and has never been afraid to talk in public about the rational and irrational consequences of the evidence.

Indeed, he was directly involved in ufology, according to the magazine *Flying Saucer Review* – which at the time when Professor Davies was associated had an unrivalled reputation throughout the world as sober and scientific, despite that somewhat anachronistic title. It reports that he had occasionally sat in on quiet seminars which they organized as a sort of brains trust among interested British academics.

The late Dr J. Allen Hynek, who is widely regarded as the father of scientific UFO research, often commented to Jenny during her meetings with him that he had a considerable regard for the input that Paul Davies had provided on the subject, usually outside of the public limelight. Hynek also took part in one 1978 TV programme put together by Jenny Randles and Peter Warrington, in which Professor Davies discussed UFOs with admirable restraint and displayed a knowledge of the data that would put to shame many media-touted UFO pundits.

So what are his considered opinions about this much debated topic?

'These stories and reports are awfully intriguing to read. I should say right away that I am not inclined just to dismiss it all as lies or hoaxes. Nor am I inclined to accept it at face value as evidence for extraterrestrial intelligence.

'I suspect that this field is more likely to be explained as some bizarre psychological experience – something to do with the mind and emerging from the depths of our psyche. I regard most of the people who claim such experiences as very sincere, but I do not share their interpretation of what has happened to them.

'I suspect that there may be no real difference between those who claim they were snatched on their way home from work by an extraterrestrial intelligence and St Paul on the road to Damascus. I simply do not think it likely there is an extraterrestrial dimension to this at all. Indeed, the more I think about the UFO subject, the more I feel convinced that it is less concerned with extraterrestrial life and more like a religious experience.'

The Evidence

15
Signals from Space?

In the winter of 1964 Jenny Randles unwittingly detected a message from outer space.

At the time she lived in a terraced house in Manchester and was just starting senior school. The family had purchased its first-ever 'radiogram', the predecessor of the modern hi-fi systems. Its frequency range was broad enough to receive quite unusual signals. Just a mile away was the converted church where the very first 'Top of the Pops' TV programmes were made and it was possible to tune the machine into the rehearsals and hear groups like The Beatles give live performances.

Twiddling became both an obsession and an art, but the strangest message ever detected was one that repeatedly spoke in a weird electronic voice and sounded something like a robot out of a science-fiction film! The noise modulated considerably, but the beginning and end words were clear enough. The voice spoke of being a test transmission and ended with the unmistakable sentence: 'This station is situated in outer space.'

At first it was considered a joke. Then it was assumed to be an early telecommunications satellite transmission. It was not continuous in transmission but it was heard frequently. It was eventually recorded on the old lumbering Grundig tape machine Jenny used. Although, of course, that tape has long since disappeared, many people heard the message and agreed about its contents. For a few months it was subject to much debate, then forgotten about.

In those days Jenny had other matters of more pressing concern to contend with, and had no idea that the space message had become the topic of even wider debate than among her family and friends. Indeed, it was only during research for this book that we learnt the full story. We obtained some old issues of Flying Saucer Review magazine and found that the 'space message' had once been the talk of the UFO field.

The first entry was in the May-June 1964 issue. It reported how the ITV station Granada carried the story on its news bulletins because Eric Lowe, a radio amateur from Wigan, had picked up 'weak signals from outer space'. It was on the short wave (which is where Jenny recalls she detected it on the radiogram) and, according to the Wigan Evening Post of 10 January 1964, 'radio experts were today trying to trace the source'.

Reportedly, the message first appeared between 12.50 and 1.10 am on 4 January and (just as Jenny ultimately concluded) was presumed to be from a brand-new communications satellite. The recording by Eric Lowe showed that the repeated message was: 'This is a test transmission for circuit adjustment purposes from a radio station of the domanial telecommunication or dorevation. This station is situated in outer space.'

Jenny notes that the middle words were rather difficult to interpret exactly but were something like this. However, the final phrase was clear and identical but had a very peculiar tone.

What makes this case so interesting is that it appears the solution was not as straightforward as imagined. There were satellites in orbit, although very few in those early years of the space age. Nevertheless, the people at Jodrell Bank refuted this explanation. They told the BBC that they had heard of the message but they had not picked it up themselves. They concluded, 'Someone seems to be doing this as a hoax. It should really be reported to the GPO [forerunner of British Telecom].'

It seems that neither Jodrell Bank nor the BBC succeeded in finding a hoaxer, even though the messages continued for at least a year (initially at a frequency of 10 MHz). Some radio amateurs noted traces of a Doppler effect, which implied the source was indeed outer space!

The second mention in *Flying Saucer Review* was a letter from a Mr S. Bagnall of Barking, Essex. He said that he had also picked up the message. He translated the middle part of 'a radio station of the global telecommunications organization' and maintained that there might be a cover-up because the government intended to use secret satellites to control nuclear missiles.

The net spread even wider in the November-December 1964 issue when the anonymous 'C.D.' from Bordeaux in France wrote to say that he/she had picked up the message from space on 19.3 MHz, although on 9 September it had switched to 9.914 MHz. The words 'Cosmos' and 'telephonicos' were heard in a conversation in an unknown language after the message. *Flying Saucer Review*'s editor, Charles Bowen, noted that these words suggested Greek.

Then came the final word (to our knowledge) in the saga. The March-April 1965 issue of *Flying Saucer Review*, featured Gordon Lindsay, who wrote from Glasgow to say that he had heard the message and used a 'beat frequency oscillator' to fill out the distortions. Upon doing so, he claimed that it magically transformed into: 'This is a test transmission for circuit adjustment purposes from a radio station of the Hellenic telecommunications organization. This station is situated in Athens, Greece.'

Thus, it was alleged, the mystery had ended; not that anyone who heard the message in 1964 or 1965 is likely to be convinced. As Jenny says, 'I would have taken an oath in court that the words were "outer space" not "Athens, Greece" . . . but then, who can say?'

We relate this story not because it represents obvious evidence for alien communication, but because it illustrates the difficulty of evaluating even the most clear-cut and simple transmissions. Much of the rest of the evidence we have to consider is less repeatable or easy to detect than this early message from outer space (perhaps via Greece) proved to be. Of course, as equipment has become more sophisticated so the temptation for hoaxers has become all the more evident.

A few years ago we investigated a couple of cases from the Sale and Altrincham areas of Cheshire where telephone callers found their lines suddenly intruded by what purported to be a secret radio message reporting an alien invasion of the north-west of England. One of these involved a lengthy ground-to-air communication from aircraft in a battle with a spaceship that ultimately crashed in Yorkshire. A couple of RAF jets bit the dust too.

Of course, this was a hoaxer circumventing the normal phone lines. However, we had no doubt that the telephone callers were innocent parties and ultimately British Telecom suspected an 'inside job', although it never revealed if the culprit had been traced.

About the same time, TV viewers in a large part of southern England had their signal blocked out for a brief time by what purported to be an alien communicating a message of peace. This was a very sophisticated hoax with the jokers tapping into the transmitters. It has since provoked numerous parodies in comedy sketches and the like and it seems improbable this will be the last or even the most elaborate attempt to con people with alien contacts.

Indeed, there are some who suggest that the strange circles appearing in cereal fields every summer (again mostly in southern England) might be part of a long-term alien-communication hoax. Speculation about the meaning of the patterns and the origin of the intelligence behind them has already filled the pages of many newspapers and UFO publications, including the ever-reliable *Flying Saucer Review*. However, occasionally the source of a message is more puzzling.

In 1953 some of the few TV viewers then in England briefly picked up a very odd test card on their screens. It identified the station as KLEE – from Houston, Texas. Nobody knew how this occurred and the mystery was compounded even further because in 1953 KLEE had been off the air and transmitting nothing for three years! Normal detection of a signal such as this would be impossible due to basic physics and the curvature of the earth. Unless it was a hoax, then the only explanation that made sense to the baffled experts was that KLEE's signal had travelled out into space, hit something about 1.5 light-years away and rebounded straight back to earth!

This may seem superficially attractive, but it has big problems. Would the power of this primitive TV station be sufficient for such a huge journey? The likely answer is no. But more importantly, for it to have hit the earth again on the rebound would be a feat statistically so improbable that it can be ruled out. Unless, of course, some alien intelligence picked the signal up, boosted it and deliberately fired it back at us. That theory gained much support among those who wanted to believe in such matters.

However, TV engineers who remembered the story told us that they thought it was a joke played by a specialist 'in the business'. In fact, it was veteran SETI scientist Frank Drake who got to the bottom of the mystery. Apparently a group of TV specialists were hoaxing people into believing they could supply them with TV equipment capable of picking up foreign stations. The foreign test cards were transmitted from much nearer home.

Will the same answer apply to the extraordinary case of the liner *QE2*? Unlike TV signals, freak transmission of radio frequencies across huge distances can occur when conditions are right in the ionsphere. However, that does not account for transmission across time! In 1978 Alan Holmes, radio officer aboard the *QE2*, picked up a Morse-code message using his ship's call-sign, Golf Bravo Tango Tango. He recognized that this was in a disused code and because others heard it too, he translated it. All the message turned out to be was a routine relay from ship to shore, noting its position. However, it was purporting to come from the previous owner of the Golf Bravo Tango Tango call-sign – the *Queen Mary*. In 1978 this old-timer was at Long Beach, California, permanently moored as a museum and a hotel.

It seemed that the message – if genuine – had to date from 1967 or before. In other words, the radio officer speculated, it had bounced off something in space at least 5 or 6 light-years out and returned home.

Needless to say there was much debate about this, especially in the pages of the telecommunications magazine with the delightfully appropriate name – *Hello World*. Again the spectre of aliens bouncing the signal back was raised, but in this case it seems even more certain that this was not so and this event had to be a hoax, although motives and culprits are hard to fathom out. Aside from the odds of a weak Morse-code message travelling all that way through space and back to earth, it would be beyond the bounds of credibility that it would only be detected by the very ship which had taken over the call-sign of the originator. Not even very clever aliens could have bounced the call back from light-years away and been able to pick out the right ship at the right moment.

The most famous radio message from space is equally problematic. This was actually picked up over half a century ago, but it was many years before its meaning was supposedly decoded.

In April 1928 Professor Carl Stormer from Norway was tuning into an experimental radio station (PCJJ, later Radio Hilversum) which was run by the Phillips company at Eindhoven in Holland. What most puzzled him was a ghost signal or echo that came some three seconds after the main one. He knew that radio waves could bounce off the ionosphere but because they travelled at the incredible speed of light (186,000 miles/ 299,274 kilometres per second), it only took a fraction of a second even to go right round the earth.

Three seconds was an impossible echo time. Even in 1969, when man landed on the moon, the delay was less than that for the radio messages to come back down to earth. The dilemma about the PCJJ echoes was that they seemed to be bouncing off something in space which was out beyond the moon.

Much discussion followed this discovery. We were still in the early days of radio and it was considered possible that some freak effect in the ionosphere, not yet fully understood, would be found to be responsible. However, Stormer notified others and they picked up the echoes as well.

On 11 October 1928, things took a dramatic new turn when there was a wave of signals with the echo no longer constant, but fluctuating between 3 and 15 seconds. This was even more ridiculous, but both Stormer and another researcher, Jorgen Hals, heard them and made a written record of the anomaly on this wavelength of 31.4 metres. Between then and April 1929 several bursts of fluctuating echo signals were picked up in both Norway and Holland, although most of the time the PCJJ signal remained normal. On 19 February 1929 the mystery had even been detected by Sir Edward Appleton at King's College in London. It was last noted by a French team in May 1929.

Between 1947 and 1949 a systematic scan of the 31.4-metre wavelength was conducted in a new effort to bring advanced knowledge to bear on the problem, but there were now so many radio stations and other transmissions flooding the airwaves that it proved impossible to tune out interference to have the clarity required. No further PCJJ echoes were ever detected, so the whole story of these unexplained signals was filed away and forgotten about.

It was raised again in the 28 May 1960 issue of *Nature*, when one of the new breed of radio astronomers at Stanford University, Professor Ronald Bracewell, suggested that we might be able to detect a message from an alien probe sent across the light-years in order to explore our solar system. It was possible to imagine that this might be programmed to stop here and to establish contact by reflecting our own radio signals back as soon as our technology became sufficiently advanced so that our society began to produce energy in the radio spectrum. In this way the probe would start announcing its presence in our solar system by puzzling radio echoes of our own messages, which, as Bracewell said, 'were reported 30 years ago . . . and never explained.'

This idea lay dormant for a while until a young Scottish astronomer, Duncan Lunan, found the notes made by Stormer and others. He used the logic Bracewell had suggested and decided to see if a message from a 'parked' space probe could be read into the fluctuating sequence of echoes. Lunan became increasingly excited (some later said he got understandably carried away) when he started using mathematical graphs to plot the echo patterns.

He ranged the sequence number of the echo (i.e. 1st, 2nd, 3rd, etc.) against the delay in seconds. Staring out from the resultant graph was what looked to him like the map of a star system. One point stood to the left on its own – as if singled out. When he reversed the plot of this point, it transferred into the main constellation as if completing the pattern.

Lunan gave lectures, TV appearances and published a book, *Man and the Stars*, in which he concluded that the message from the 'probe' was a star map of its home system. The singled-out point was the star system from which the probe had set out – Epsilon Bootis, a fairly distant system where life was possible, although the conditions were hardly ideal.

Predictably the critics were soon on the attack. Lunan was accused of 'manipulating data', which depends on how you interpret what he did. For instance, his map had one major anomaly: the star Arcturus was not in the right place, so he concluded that if you juggled about with its motion through space you could get it to more or less fit the map from its position 13,000 years ago – although it is unclear to us whether other stars would have altered position too. So Lunan concluded that the probe's map dated back to when it was first launched at our solar system.

The most thorough analysis was carried out by astronomer and engineer Anthony Lawton, who concluded in his 1975 book, co-written with Jack Stoneley, *Is Anyone Out There?* that there were acceptable natural explanations for the echoes – such as theoretical plasma clouds in the ionosphere which could slow down a radio signal before it reflected.

The most important discovery Lawton made was the records of the readings by Sir Edward Appleton in 1929. He tried Lunan's method of plotting them and a meaningless graph emerged, certainly not a star map of the Bootis system.

It is important to remember the crudeness of the maps in question. The original PCJJ data comprised just 14 echoes, so the map is formed from very few points. In addition, half of the echoes were of eight seconds duration. When you plot the graph (or map), these form a solid line down the middle. Lunan interprets only the other seven points as being stars and this vertical line as a barrier to separate the star map to the right from the singled-out 'home' star to the left. In other words, the star map of Bootis is formed from just 6 out of the 14 recorded echoes, and subsequent patterns recorded elsewhere on the PCJJ frequency were never identical to it.

Duncan Lunan did endeavour to justify this reasoning by suggesting that once we had failed to prove our intelligence to the probe (for example, by sending the star map picture back to it), it gave up reflecting the echoes. Either it switched off to await another sign of intelligence or it was programmed to try other methods. Perhaps the long delayed TV and ship-to-shore messages were a result of these further probe experiments.

To be fair to Lunan, he saw the problems of his theory and it was honestly offered as an exercise in deductive thinking. Presently he is an active member of the British science-fiction community, often contributing articles on astronomical science to its magazines. His evaluation of the PCJJ radio echoes may not have revolutionized the world of cosmology, but it was certainly a fascinating concept – real life science fiction.

Whether he simply allowed his imagination to go too far or did perceive an amazing truth, may never be resolved. However, the basic principle behind the star probe remains valid. After all, we have already sent our own less-sophisticated machines into the far reaches of space where they may one day encounter an inhabited planet.

Periodically, we still hear stories that radio messages have been intercepted from another world. When the incredibly distant signals of utterly fantastic energy that we now call quasars (for quasi-stellar sources) were detected, there was a time when some scientists thought they might be alien beacons. Pulsars, which are throbbing lighthouses of power, are natural energy sources too, but the idea that they were artificial crossed more than a few minds until they were better understood.

More recently, when an anomalous signal was picked up during a search programme by the radio telescope system Cyclops (so named because of its many 'eyes' acted as one), it made the *New York Times* with the sedate and rather restrained story and headline: MYSTERIOUS TRANSMISSIONS CONTINUE TO BAFFLE ASTRONOMERS.

The signal, detected on 14 February 1976, was a more sophisticated form of the PCJJ echoes; a complex blend of fluctuating tones with a 'rhythmic structure' that included many sounds well beyond the threshold of human hearing. The baffled project director at the remote Arizona site, Dr John Oliver, said that all efforts to translate it had failed since it bore no resemblance to human speech. Indeed, Hermann Bernard (a musicology professor from the Massachusetts Institute of Technology), had translated it into musical notation. Bernard said that the result 'looked more like an expressionistic drawing than a piece of music. At times over 500 notes are sounded simultaneously, and the range of pitch appears to be infinite.'

This may have been the genesis for Steven Spielberg's idea to use musical notes as the alien communication method for his film *Close Encounters of the Third Kind*, then in early production. But surely this is what we ought to expect? An alien message would be far more likely to appear indecipherable and weird, rather than a straight reproduction of what *we* would do if positions were reversed.

This signal came from the star Ophiuchi, some 17 light-years away. It is one of the few known to have a companion which is considered likely to be the sign of a planetary system.

Yet the Cyclops scientists were baffled by the lack of interest. The American government said the programme was costing too much and began taking staff away. One senator even accused the team of fabricating the musical message to try to rescue the project and 'win public support for a scientific luxury item.'

The whole incident quietly slipped away and was never reproduced or verified. The world did not end. People did not rush into the streets screaming that the aliens had arrived. In fact, it was as if it had never happened.

However, there is one interesting fact about this 'message' that might be significant and help to explain the subdued response. We tend to believe that aliens out there will be like us, will have selected our sun as a likely home of fellow beings and will beam a message in our direction. But as we have seen, *if* this 1976 message was from another world, the aliens there do not appear to have been like us and they had evidently concluded we were not even worth checking out. For this message from Ophiuchi was bypassing earth altogether and heading straight for the star Eta Cassiopeiae some 18 light-years beyond our humble – and possibly cosmologically irrelevant – abode.

16
Was God an Astronaut?

'Have you ever thought,' Captain Brent asked me quietly, 'that the earth might be a colony started by another world?'

I looked at him, startled. 'I've heard it suggested, but – do you actually believe it?'

'I'm certain of this much,' replied Brent. 'A race far more technically advanced than we are today was on earth thousands of years ago.'

He swung around to a cabinet, took out a folder. 'The Hydrographic Office of the Navy has verified an ancient chart – it's called the Piri Re'is map – that goes back more than 5,000 years. It's so accurate only one thing could explain it – a worldwide aerial survey.'

This conversation between 'Captain Brent' and Major Donald Keyhoe was reported in Keyhoe's book *Flying Saucers: Top Secret*. Was God an astronaut? This was the question posed by the *Sunday Mirror* in Britain, when, in 1968, it serialized *Chariots of the Gods?* by its then unknown Swiss author, Erich von Dänekin. But the book, along with its successors, was to become an international best seller, and von Dänekin, a household name. Ironically, despite the book's controversial claims, it does *not* state that God was an astronaut! However, it did profess to present archaeological evidence that the earth was once the watering-hole for extraterrestrial playboys. Not only that, but von Dänekin claimed that a deliberate programme of genocide was carried out on the human race, leading to a new refined version of *Homo sapiens*. Proof of all this, it was claimed, was to be found in the Bible and other ancient texts, and various artefacts scattered about the world.

During his explorations, von Dänekin found what he decided were 'airstrips' laid out on the Peruvian plains, gigantic stone blocks carved to perfection by 'alien technological know-how', 'petrified excrement, possibly not of human origin', electric dry batteries in Baghdad, an aluminium belt in China, a non-rusting iron pillar in Delhi, cut crystal lenses in Egypt and Iraq, and cave drawings of space visitors. All of this was proof, the ebullient Swiss author claimed, that technological wonders existed in pre-history, and that they were gifts from the gods.

But von Dänekin was not the only author to make such claims, nor was he the first, although he was certainly the first to write a best seller on the subject! The Desmond Leslie part of *Flying Saucers Have Landed*, first published in 1953, provided most of the sources for successive writers. Other books and authors included *UFO and the Bible* by M. K. Jessup (published in 1956), *The Sky People* by Brinsley Le Poer Trench (1960), *Gods or Spacemen* and *Spacemen in the Ancient East* by W. Raymond Drake (1964), *Flying Saucers and the Scriptures* by John W. Dean (1964), and *God Drives a Flying Saucer*, by R. L. Dione (1969).

There were many more, and some of them were reprinted in the wake of von Dänekin's success. Some, though not all, were factually inaccurate and unscholarly. But they were aimed at a particular market and volume book sales resulted.

Not all of these writers were merely out to earn a fast buck. Several, such as Barry H. Downing (who wrote *The Bible and Flying Saucers*), a pastor with a bachelor of arts degree from Hartwick College, Oneonta, New York, where he majored in physics, had actually carried out some extensive research. In the mid-1970s, we attended a lecture in Manchester given by the late W. Raymond Drake. His lecture, like his books, was very detailed indeed.

Extraordinary claims require extraordinary evidence, but in the short term this did not seem to matter. Von Dänekin and his peers had convinced the general public that their interpretation of biblical events were as valid as those of religious leaders, orthodox historians and archaeologists. And why not? Many episodes in the Old Testament can be seen in terms of a period of history where extraterrestrial visitors took it upon themselves to meddle with the genetic and cultural evolution of early humans. Russian ethnologist M. M. Agrest proposed a connection between some biblical events and alien visits in 1959.

Apart from this, the Ancient Astronaut theorists had successfully tapped into a raw nerve. The public loves to see stuffy scientists pulled down a peg or two! Know-alls have never been popular, and scientific 'facts' and theories have been shown to be wrong many times in the past.

One of the main bones of contention in the Ancient Astronaut controversy was a passage from 1 Ezekiel, where the prophet describes a 'vision' while standing by the banks of the River Chebar:

As I looked, behold, a stormy wind came out of the north, and a great cloud, with brightness round about it, and fire flashing forth continually, and in the midst of the fire, as it were gleaming bronze. And from the midst of it came the likeness of four living creatures. And this was their appearance: they had the form of men, but each had four faces, and each of them had four wings. Their legs were straight, and the soles of their feet were like the sole of a calf's foot; and they sparkled like burnished bronze. Under their wings on their four sides they had human hands. And the four had their faces and their wings thus: their wings touched one another; they went every one straight forward, without turning as they went. As for the likeness of their faces, each had the face of a man in front, the four had the face of a lion on the right side, the four had the face of an ox on the left side, and the four had the face of an eagle on the back. And their wings were spread out above; each creature had two wings, each of which touched the wing of another, while two covered their bodies. In the midst of the living creatures there was something that looked like burning coals of fire, like torches moving to and fro; and the fire was bright, and out of the fire went forth lightning. And the living creatures darted to and fro, like a flash of lightning.

Now as I looked at the living creatures, I saw a wheel upon the earth beside the living creatures, one for each of the four of them as for the appearance of the wheels and their construction: their appearance was like the gleaming of a chrysolite; and the four had the same likeness, their construction being as it were a wheel within a wheel. When they went, they went in any of their four directions without turning as they went. The four wheels had rims and they had spokes; and their rims were full of eyes round about. And when the living creatures went, the wheels went beside them; and when the living creatures rose from the earth, the wheels rose.

Over the heads of the living creatures there was the likeness of a firmament, shining like crystal, spread out above their heads. And under the firmament their wings were stretched out straight, one toward another; and each creature had two wings covering its body. And when they went, I heard the sound of their wings like the sound of many waters, like the thunder of the Almighty, a sound of tumult like the sound of a host; when they stood still, they let down their wings.

Ezekiel then saw a 'vision' of God seated on a throne, who instructed him on a programme of reform. After this, the bemused man was lifted up and transported to the place where he was to start his work:

Then the spirit lifted me up, and as the glory of the Lord arose from its place, I heard behind me the sound of a great earthquake; it was the sound of the wings of the living creatures as they touched one another, and the sound of the wheels beside them, that sounded like a great earthquake.

By any standards this is an enigmatic piece of writing, open to interpretation. There seems no reason for Ezekiel inventing the story, in the way that Christ told stories which served as metaphors for deeper teachings. Could it have been a vision in the sense of an hallucination brought about by mind-expanding drugs? Yet what comes across more than anything in this passage is the utter complexity of the vision. Ezekiel's frustration is barely masked as he struggles to convey to the reader something totally outside the experience of early biblical humans.

In *Chariots of the Gods?*, Erich von Däniken has no doubts about the source of Ezekiel's 'vision' – nuts-and-bolts flying machines of extraterrestrial origin:

Ezekiel gives precise details of the landing of this vehicle. He describes a craft that comes from the north, emitting rays and gleaming and raising a gigantic cloud of desert sand. Now the God of the Old Testament was supposed to be omnipotent. Then why does this almighty God have to come hurtling up from a particular direction? Cannot he be anywhere he wants without all this noise and fuss?

In addition to his precise description of the vehicle, Ezekiel also noted the noise the novel monstrosity made when it left the ground. He likens the din made by the wings and the wheels to a 'great rushing'. Surely this suggests that this is an eye-witness account?

He thought this locomotion quite incompatible with the idea of an omnipresent God. Von Dänekin received a shot in the arm from Josef Blumrich, chief of the systems layout branch at the Marshall Spaceflight Center of NASA. He worked on the Saturn rocket and the design of Skylab. He too thought that Ezekiel had witnessed the arrival of extraterrestrial hardware in the form of some sort of helicopter. His very technical thesis was expanded into a book with the publication of *The Spaceships of Ezekiel*.

Blumrich was no dummy. He knew what he was talking about, and the scientific fraternity, who had smiled condescendingly at von Dänekin and his contemporaries, actually sat up and took notice – for a time, at least. This is Josef Blumrich's interpretation of the biblical passage mentioned earlier: 'We should consider that Ezekiel first saw this vehicle at a distance of about 1,000 metres [1,100 yards]; at that moment the nuclear engine fired, probably with some white clouds of condensation.'

The rounded 'soles of the feet', as described by the prophet, was the very design Blumrich and his colleagues had used to allow the legs of a spacecraft to slide on landing. Blumrich remarked on how unbird-like the wings of the 'creatures' sounded. He saw them as the blades of helicopter units. As for the four faces of the creatures, Blumrich equates this with our own pilots who paint the faces of birds and animals on to the noses and fuselages of aircraft.

More interesting, however, is Ezekiel's repetitive and detailed descriptions of the 'wheels', and the effect they had on this twentieth-century space engineer.

'No one,' Blumrich wrote, 'has ever taken seriously the functional description which indicates that the wheels could move in any direction without being turned or steered.'

Blumrich *did* take it seriously. He designed the Omni-directional Wheel and patented the idea in 1974. From the biblical description, he went on to reconstruct a humming-top-shaped spacecraft which would accommodate four helicopter landing vehicles. He did everything but construct a nuclear-powered engine to allow the craft to be built and tested!

But, as Edward Ashpole, author of *The Search For Extraterrestrial Intelligence*, pondered: does this spaceship belong to Ezekiel or Blumrich? Arch-UFO-debunker, astronomer Donald Menzel, postulated a meteorological explanation for Ezekiel's 'vision'. But most sceptics were more cautious and they were kinder to people like Menzel than they were to von Dänekin. The chinks in his armour became gaping holes as the sceptical camp took up the challenge. The honeymoon was over.

Two books appeared in answer to von Dänekin's claims, *Crash Go the Chariots* by Dr Clifford Wilson and *The Space Gods Revealed* by Ronald Story. Clifford Wilson, an Australian, was later to become fascinated by the UFO phenomenon and to write an intriguing book called *UFOs and Their Mission Impossible*. Story, a confirmed sceptic of the Ancient Astronaut theory, took von Dänekin to task and tore his ideas to pieces.

There were extreme charges made, some of which von Dänekin vigorously denied; not that the dispute hindered his career. In excess of 40 million copies of *Chariots of the Gods?*, *Return to the Stars*, *Gold of the Gods*, *In Search of Ancient Gods* and *Miracles of the Gods* have been sold world-wide. And in 1989 *Chariots* was again reprinted in paperback!

At his worst, von Dänekin wrote of things which could not be independently verified and on occasion was accused of having misrepresented the stories of others as personal experiences of his own. A prime example of this was *Gold of the Gods*. At the start of this book, von Dänekin states:

To me this is the most incredible, fantastic story of the century. It could easily have come straight from the realms of Science Fiction if I had not seen and photographed the incredible truth in person.

What I saw was not the product of dreams or imagination, it was real and tangible.

A gigantic system of tunnels, thousands of miles in length and built by unknown constructors at some unknown date, lies hidden deep below the South American continent. Hundreds of miles of underground passages have already been explored and measured in Ecuador and Peru.

An explorer called Juan Moricz allegedly discovered the caves in 1965. Von Dänekin went to see him and, according to his book, was given a conducted tour of the passages, where he examined some of the many thousands of gold artefacts said to be hidden there:

The entrance, cut in the rock and wide as a barn door, is situated in the province of Morona Santiago. Suddenly, from one step to another, broad daylight changed to pitch darkness. We switched on our torches and the lamps on our helmets, and there in front of us was the gaping hole which led down into the depths. We slid down a rope to the first platform 250 feet [76 metres] below the surface. From there we made two further descents of 250 feet. Then our visit to the age-old underworld of a strange unknown race really began.

Then they entered a huge hall which boasted a table and seven chairs, 'as hard and heavy as steel' but manufactured from some sort of plastic. There von Dänekin described finding the images of animals behind the chairs: saurians, elephants, lions, crocodiles, jaguars, camels, bears, monkeys, bison and wolves, with snails and crabs crawling between them. Von Dänekin wrote that the whole thing was like a zoo, and that the animals were made of solid metal of some kind.

There was more – much more! A 'library of metal plaques', a human skeleton carved from stone, a stone dome, carved figures wearing 'space gear' and ... gold were discovered. Von Dänekin claimed he was prevented from photographing much of what lay in the underground labyrinth by his companion, afraid of destroying the 'magic' which resided there. Nor could a sample be procured to convince the outside world. If any were levered from a pile, he was allegedly told, there was a chance an avalanche might start!

Some commentators fear that *Gold of the Gods* was a serious error of judgement by Erich von Dänekin. It transpired that the best-selling author had never been in any Ecuadorian underground caves. The allegations were made in an interview with Moricz in the West German news magazine *Der Spiegel* of 19 March 1973.

In it Moricz claimed that von Dänekin had never been in the caves, and had 'squeezed' the story out of him over a period of days. The information was the results of Moricz's own expeditions through the labyrinth. He said that von Dänekin had promised only to refer to the material in the context of a book Moricz was writing.

This rock painting is one of many which date back to around 6000 BC. Von Dänekin thinks it represents an alien in a spacesuit, but Ronald Story believes it is a man wearing a ritual mask and costume. (*G. Franceschi*)

The caves themselves, allegedly running under the Andes from Ecuador to Peru and Bolivia, proved as illusive as von Dänekin's journey through them. Professor Fritz Stibane of the University of Giessen, in West Germany, found no underground-tunnel system and no gold artefacts. The Ecuadorian government, four years before, had also checked out similar cave rumours, but to no avail. Even Moricz now claims he had not visited this vast cave system, but based many of his statements on 'mythology'. A local priest, Father Crespi, according to von Dänekin, possessed a collection of aretfacts 'priceless for its weight in gold', lying around in his backyard! However, according to investigators of the story, Crespi was an eccentric who traded clothes and small sums of money for the artefacts made by local Indians.

Erich von Dänekin, to his credit, admitted in an interview in *Playboy*, in August 1974, that Moricz had denied taking him into the caves and showing him the metal objects, but he was adamant that whatever this man had said, he *had* visited some caves, although 'not at the main entrance.'

When interviewed for the BBC as part of a *Horizon* documentary made in 1978, he told the reporter that in Germany it was accepted practice to present someone else's story as one's own. But the truth was that readers who did not accept his explanation must have felt badly let down.

A lesser bone of contention was the Piri Re'is Map. This is, in fact, a fragment of a map which was in the possession of Admiral Piri Re'is of the Turkish Navy, and discovered in 1929 during conversion work at the Palace of Topkapi in Istanbul. Dated 1513, it showed South America, part of Africa and an area of coastline now buried beneath the Antarctic.

In *Chariots of the Gods*, Erich von Dänekin said:

All these maps were handed over to the American cartographer Arlington H. Mallery for examination. Mallery confirmed the remarkable fact that all the geographical data were present, but were not drawn in the right places. He sought the help of Mr Walters, cartographer in the Navy Hydrographic Bureau. Mallery and Walters constructed a grid and transferred the maps to a modern globe. The maps were absolutely accurate – and not only as regards the Mediterranean and the Dead Sea. The coasts of North and South America and even the contours of the Antarctic were also precisely delineated on Piri Re'is' maps. The maps not only reproduced the outlines of the continents, but also showed the topography of the interiors! Mountain ranges, mountain peaks, islands, rivers and plateaux were drawn in with extreme accuracy.

Von Dänekın maintained that the original map, which for some strange reason he referred to in the plural, could only have been produced from pictures taken in outer space. The elongated shape of the

The Piri Re'is map discovered in the eighteenth century, and said to date even further back. Although Story used it to help debunk Erich von Dänekin, the map still represents a true mystery. (*University of Arizona Library*)

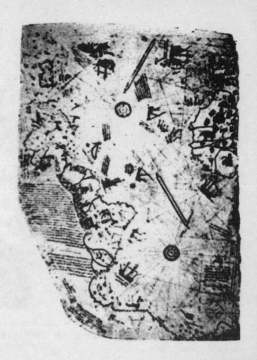

continents were identical to the distortions evident on our own satellite photos. In support of this, he cited Professor Charles H. Hapgood and mathematician Richard W. Strachan. But this further 'proof' of alien visitations also did not impress von Dänekin's critics.

Ronald Story states that the map is quite *in*accurate. He cites Hapgood as showing that, compared to a modern map, there is very little resemblance between Cuba, Crete and the western Mediterranean. Nine hundred miles (1,450 kilometres) of South-American coastline are missing, and the River Amazon appears twice.

As usual, von Dänekin was vociferous in his support for an extraterrestrial source for the mystery, but he was making the same errors that he accused sceptical archaeologists and historians of making – selectively using facts to fit his theory. Yet the Piri Re'is map is undoubtedly mysterious. Story makes this observation in *The Space Gods Revealed*: 'The Piri Re'is map, although not 'proof' of extraterrestrial visitations, remains a fascinating subject for study, especially when considered along with other old maps that show a continent suggestive of Antarctica.'

One cannot fail to be impressed by this sixteenth-century relic, despite its inaccuracies. Was it a copy of a much earlier map? Were the

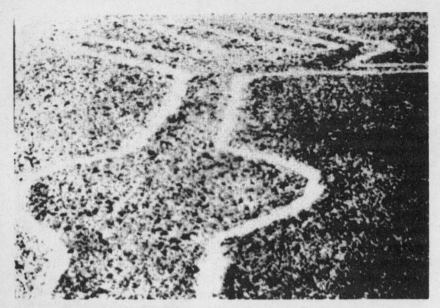

One of the photographs that caused von Dänekin some trouble. He said the lines reminded him of aircraft parking bays – yet they were only yards across! (*Marcel Homet*)

inaccuracies incurred during copying, or is the original so old it was made at a time when the continents and islands were in slightly different positions from the one's occupied today? However, it has to be said that those elongated Americas do remind one of how they would appear from a fixed point, staring down from space.

Von Dänekin dealt himself another joker over the Nazca lines – an area of land between Nazca and Palpa, in southern Peru, measuring 40 miles by 9 (64 by 15 kilometres), literally covered with gigantic drawings of animals, birds and geometrical figures executed with amazing precision. The most astonishing thing about these drawing is that they are only obvious from the air.

Again, von Dänekin claimed that the animal designs were made by the Incas as landmarks for visiting spacecraft. Of the hundreds of parellel lines and other geometrical shapes, he said, after overflying the area, that they reminded him of a gigantic airfield. The critics lost no time in exposing flaws in his speculations. In *Chariots of the Gods?*, von Dänekin reproduced a photograph with the comment: 'Another of the strange markings on the Plain of Nazca. This is very reminiscent of the aircraft parking bays on a modern airport.'

Indeed it is, until one appreciates that the 'bays' are in fact the claws of a condor, and just a few yards across! Von Dänekin accepted that the idea was absurd in the previously mentioned BBC documentary. If the plain *was* a huge airport, why is there no trace at all of any buildings? Why are there no unambiguous alien artefacts in this huge area?

The lines, created by removing the surface stones and scratching away at the top soil, were the subject of intense and dedicated investigation by German mathematician Maria Reiche. Up until her recent death, Ms Reiche spent almost 50 years living and breathing this ancient Inca mystery. For much of that time she was close to poverty, but the miles of parallel lines, animal drawings and finely executed geometrical designs would not let her go.

With American Professor of History Paul Kosok, Reiche devised an astronomical explanation for the Nazca phenomenon. Some of the animal designs seemed to coincide with the ancient star constellations in the southern hemisphere, and many of the straight lines which criss-crossed the plain appeared to point towards the sun, moon and various stars. As the climate was pretty much the same all year round, Reiche thought it likely the drawings were some sort of astronomical calendar to help the Nazca farmers know when to sow crops and to determine when the rivers would fill from the distant rainfall.

Dr Gerald S. Hawkins, author of *Stonehenge Decoded*, decided to put the theory to the test, using a complicated computer program. The results revealed that Nazca had little astronomical significance. This did not please Maria Reiche, who hotly contested the findings.

None of this answered the questions of *how* the drawings were made with such accuracy, or for what reason, and why they were produced to be visible properly only from the air?

American businessman Jim Woodman reckoned he knew some of the answers. The Nazca draughtsmen were helped in their work by instructions relayed from hot-air balloons hovering above the plains. The balloonists might also have used the completed designs as beacons for landing, much as von Dänekin's extraterrestrials would have done.

But why so many 'beacons'? The Peruvian legends of flight in Inca times, interpreted by the Ancient Astronaut camp as proof of space flight, were now seen in terms of hot-air ballooning. To prove the point, Woodman and British balloonist Julian Nott flew a balloon over Nazca. It was manufactured from materials which could have been found locally. The gondola basket was made of totora reeds from the shores of Lake Titicaca, and joined to the balloon by ropes plaited from locally grown fibres. The envelope was sewn from a finely woven fabric similar to cloth, also found in the Nazca area.

After flying for two minutes, the primitive balloon had reached a height of almost 400 feet (122 metres), before crashing back to earth.

The men jumped to safety just before the unpredictable flying device took off again.

Did balloonists play a part in the creation of the Nazca lines? Maria Reiche thought it more likely the drawings were extrapolations of smaller models. Were the lines pathways to family shrines, as suggested by American anthropologist William Isbell? Thousands of items of pottery have been recovered over the years.

Actually, there is no convincing explanation, but many quite rational suggestions. Ancient Astronauts may be as valid an argument as hot-air balloons, although far more contentious. The area is so vast, the drawings so accurate, that it must have taken the skills and manpower of whole tribes over a long period of time to complete the task. And with such dedication, there must have been a very good reason indeed to carry out the work in the first place.

There has been a duality in the role that Erich von Dänekin has played in the Ancient Astronaut controversy. Many before him made similar claims, but he made the sort of impact which brought the subject to the attention of the world. As in any publishing venture, there was an element of luck. Nobody can sit down and predict a best seller at the time of writing. *Chariots of the Gods?* came at a time when the first human (as far as we know!) was about to land on the moon. The whole world was caught on a wave of euphoria. The sky was no longer the limit. Manned journeys to other worlds were now possible, and happening. The public consciousness was ripe to hear that alien astronauts of a former time had visited us.

Yet while we can thank von Dänekin for that, his repetition of what we must call 'tall tales' played straight into the hands of the sceptics, fogging the real issues.

Was God an astronaut? Despite all this, von Dänekin did have some supporters with sound scentific backgrounds. Apart from Josef Blumrich and Barry H. Downing, mentioned earlier, there was also Swiss electronics engineer Heinrich Gosswiler. He saw an ancient cave drawing in Santa Barbara, California, in terms of a complex scientific diagram.

If Erich von Dänekin suffered for a lack of scientific credibility, then Robert K. G. Temple did not. After nine years of intensive and original research, Temple produced *The Sirius Mystery* in 1976. It presented very convincing evidence that the ancient Egyptians had been·visited between seven and ten thousand years ago by beings from the star system Sirius. It was a scholarly work and anthropologically sound. Despite his thesis of alien contact, he was embarrassed in case he was dumped in the same pot as the Ancient Astronaut crowd.

Robert Temple's journey of discovery began with the reading of an essay by two eminent anthropologists, Marcel Griaule and Germaine Dieterlen. It concerned the cosmological beliefs of the Dogon people of West Africa. One paragraph in particular impressed him:

The starting point of creation is the star which revolves around Sirius and is actually named 'Digitaria star'; it is regarded by the Dogon as the smallest and heaviest of all the stars; it contains the germs of all things. Its movement on its own axis and around Sirius upholds all creation in space. We shall see that its orbit determines the calendar.

This African people, which Temple later grew to believe was derived from the ancient Egyptians, seemed to possess a profound knowledge of the Sirius system. It knew that Sirius, the brightest star in the sky, had a companion, which was much smaller than Sirius 'A', and very 'heavy'. It also knew that Sirius 'B' took 50 years to orbit its larger companion, even though this smaller star is visible only through the most powerful telescopes.

The astronomer Johann Friedrich Bessel speculated in 1844 that Sirius was a binary star system, but its existence was not confirmed until 1862 by Alvin Clark. Over half a century later, its eccentric properties were determined, and out of that came the discovery of the first 'white dwarf'. Sirius 'B', although only three times the radius of the earth, has a mass just a little less than our sun.

The more Robert Temple investigated, the more astounding were his discoveries. The Dogon people had a whole range of astronomical knowledge which seemed to have been handed down over thousands of years. It knew that planets moved in elliptical orbits, and it referred to the moon as 'dry and dead like dry dead blood'. It was aware of Saturn's rings, that the earth turned on its axis, and the four major moons of Jupiter.

Temple found an entire mythology which matched his scientific data. This knowledge, and much more, had been brought to the earth by the 'Nommos' – an amphibious race of aliens from the Sirius star system. This is how the Dogon describe the landing of the Nommos 'ark' – presumably a spacecraft:

The ark landed and displaced a pile of dust raised by the whirlwind it caused. The violence of the impact roughened the ground. He is like a flame that went out when he touched the earth.

The 'ark' then moved to a hollow which filled with water, whereupon the aliens emerged. According to Berossus, a Babylonian historian who lived at the time of Alexander the Great, 356–23 BC, this is how the Nommos looked:

The whole body of the animal was like that of a fish; and had under a fish's head another head, and also feet below, similar to those of a man, subjoined to the fish's tail. His voice too, and language, was articulate and human; and a representation of him is preserved even to this day . . . . When the sun set, it was the custom of this Being to plunge again into the sea, and abide all night in the deep; for he was amphibious.

The comparison with a world mythology of mermen and mermaids is obvious. Temple found representations of the Nommos all over the Middle East. Stone reliefs and drawings abounded of beings adorned with fish tails. One of them, a Babylonian semi-demon called Oannes, was said to have founded the first civilizations on earth. In support of this, Temple points out how the Sumerian culture seems to have sprung up out of nowhere, and how 3,400 years before Christ, the Egyptians passed rapidly from a neolithic culture to one capable of writing and constructing huge buildings. He quotes historians and archaeologists to support his theories.

The Sirius Mystery is a very persuasive argument for the Ancient Astronaut theory. However, it does have flaws. Apart from Sirius 'A' and Sirius 'B', the Dogon maintain there is a third star. Sirius 'C' was supposedly observed by astronomers in 1920, 1926, 1928 and 1929.

Since then others have searched for it, but they have found no trace. Secondly, the Sirius system does not seem to offer the sort of stable conditions of an earth-type planet to promote intelligent life. Finally, critics have argued that we cannot be sure that the Dogon's detailed astronomical knowledge did not come from missionaries passing through.

We find this latter argument, although very relevant, the least likely to be valid. First of all, this would need to have happened after 1915 when Sirius 'B' was not only confirmed, but then recognized as a 'heavy' star. Surely our hypothetical missionaries would be more concerned in converting the Dogon towards worshipping the Christian God than giving detailed lectures on physics and astronomy? In any case, accepting such an idea means that the Dogon priests were lying about the information having been passed down through generations. The knowledge was only made available to Griaule and Dieterlen after strong mutual trust had been built up. This point about missionaries is one of the most common criticisms of Temple's thesis. However, he has been remarkably adept at attempting to answer all charges.

One of the most interesting debates arose with astronomer Carl Sagan, who attacked the research in the August 1979 issue of *Omni*. Temple has complained that he has sometimes been prevented from giving a full reply to his critics. He singles out the science journal – *Nature* – as a case in point, and claims it was reluctant to give him the space he requested.

The *Zetetic Scholar* is a serious discussion outlet on strange phenomena published by sociologist Dr Marcello Truzzi, of Eastern Michigan State University. In issue 8 (1981), Temple was able to respond to some of these points.

He noted that the supposed discovery of information about Sirius from twentieth-century missionaries is not possible, and cited the views of Dr Dieterlen:

> As she has spent most of her life living with the Dogon, and knows them and their traditions more intimately than anyone else alive, her opinion on a possible Western origin for the Sirius traditions of the Dogon is of the highest importance. She answers such suggestions with a single word: 'Absurd!'

Dr Dieterlin also has a Dogon artefact which shows the three-star system of Sirius. This has been reliably dated to be 400 years old. Furthermore, Dieterlin questioned the head of the missionaries in Mali who confirmed that none of their number contacted the Dogon prior to 1949, by which date anthropologists had already recorded details of their Sirius traditions.

Robert Temple told Dr Sagan, who was probably not convinced by it: 'I have yet to encounter a single criticism of *The Sirius Mystery* to which there was not a satisfactory reply.'

Were we visited, indeed, was our species engineered, our intellect artificially stimulated, by visitors from another world? Or in truth, can our historians and archaeologists explain it all away in mundane terms?

Von Dänekin was probably quite sincere and skilful in his efforts to prove his case, but he was undoubtedly over-zealous and stands accused by many critics of having put his case badly. Because of this, the whole subject has floundered as a scientific topic, when it may – in part – be worthy of deeper exploration.

Others, such as Temple, have established a basis for enquiry, but key questions remain unanswered. However, any objective reading of the Old Testament and other ancient texts fills one with a feeling of unease. Interference, it seems, may have come from somewhere.

17
Aliens in Orbit

At the moment our trips into space bear no relation to the journeys necessary for a proper exploration of space. We have gradually inched our way on to the front-door step of the universe by landing on the moon and Mars. To reach the nearest planet which might play host to life is an entirely different proposition. Using the door-step analogy again, it's like setting out on a solo walk to the far side of the world.

The distances we have achieved so far, great as they may seem, are nothing compared to what is required to reach other solar systems. To get there also requires a form of technology that we neither possess nor can dream of possessing, not at least in the foreseeable future.

Nevertheless, most of us have been conditioned by a generation of TV serials, like *Star Trek*, to believe that getting to other solar systems is no more difficult than reaching the moon. Indeed, many non-scientists seem to think that it's as straightforward as going on a cross-town bus! And if we can do it, then others can do it too, just as easily. So the idea that alien probes could come here has become a popular one. Inevitably the very people who might be expected to have had the best chance of seeing such extraterrestrial probes are our own astronauts. Not only have they been where the aliens are believed to be, but any intelligent species would surely want to watch our missions into space and so they would pay special attention to our astronauts.

That kind of cultural conditioning may explain why there have been persistent and widespread rumours that both Soviet and American space flights have encountered more than they bargained for and have not told all there was to tell about these things. Some contend that the hasty abandonment of lunar landings in the early 1970s was nothing to do with the incredible costs of the missions or the desire to switch funds to new projects (such as the re-usable Shuttle), but to do with the 'fact' that NASA had met the aliens.

The early days of the Soviet space programme were shrouded in military secrecy. There are rumours about cosmonauts dying on hushed-up missions hastily arranged to beat the Americans in the race for the moon. Also some see it as odd that (officially) the USSR never went to the moon at all.

One case attracting much attention at the time was that of Voskhod 1, which carried a crew of three, including Vladimir Komarov. After an unexpectedly short flight of just one day, it landed in Soviet Asia on 12 October 1964 and many science commentators were puzzled. Was there something wrong with the spacecraft? Why did the crew report that they wanted to stay up in orbit because they had seen many interesting things and wanted to investigate them more fully? The request was denied.

This statement led to intense speculation, especially among the UFO fraternity. Gordon Creighton, now editor of *Flying Saucer Review* commented in a piece titled 'Astronauts forced down by UFOs?' (May-June 1965). Author Tim Good, who feels there is a major Soviet cover-up of UFO data, reviewed many of the stories surrounding this flight in *Above Top Secret*, although he did note that the astronauts themselves never said anything that might give a hint. Indeed, in their later writings, the crew of Voskhod 1 explained their cryptic comments as a request to conduct more study of the Aurora which they observed at the South Polar regions. This seems to account more than adequately for their remarks without bringing aliens into the picture.

These escalating stories provided material for some successful books at the height of the space programme such as Don Wilson's *Our Mysterious Spaceship Moon*. He summarized his views, in 'Our Astronauts Saw UFOs', an article which appeared in July 1977 in a short-lived, newsstand magazine in America called *Official UFO*.

It is true that the millions of photographs taken in space do occasionally show odd blobs of light on them. If you stretch your mind far enough, you can visualize these as alien spacecraft, but in truth they remain nothing more than odd blobs of light.

Official explanations from NASA do seem acceptable – for example, sun reflecting off bolts on the window or pieces of junk, such as unmanned satellites, in similar orbit patterns. There have even been cases ascribed to astronauts 'boldly going where no man has gone before' and using outer space as a toilet for their UFO-like waste matter! Sunlight beyond the protective filter of an atmosphere is so strong that it does produce odd effects.

To be fair to Wilson, he does note that 'writers sometimes have gone overboard in reporting strange objects in the skies; at times even reliable reports have erred on the side of mystery.' He cites an example of the pilot Major Robert White who in July 1962 flew one of the very first missions to the edge of space aboard an X-15 rocket plane. Ufologists point to White's comment over the radio: 'There are things out there. There absolutely are!' (see *Flying Saucer Review*, September-October 1962). Indeed, even in 1987 the methodical and much respected Tim Good discussed this case in his best-selling book *Above Top Secret*.

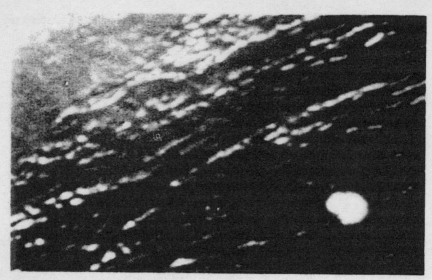

One of dozens of photographs taken by astronauts said by some to depict UFOs orbiting the earth. In this case, NASA explained that the image was caused by sunlight reflecting off a window bolt. (*NASA*)

However, as Wilson remarked, when he checked the original sources Major White did say what is reported and saw one unidentified smallish blob, but it was equally apparent that most of the things he described were small, white flaky bits of material, which he thought were fuel burn up or even ice crystals shearing off his craft. Anyone who has seen NASA film of space flights will recognize the phenomenon being referred to.

The photographs we have of alleged encounters between our astronauts and aliens are all pretty unconvincing. While ufologists usually contend that they show, for example, 'two UFOs rising out of a moon crater' (during the historic first lunar landing by Apollo 11), the official solution (light reflection off the window) seems far more plausible if you are not a committed 'believer'.

Only one photographic case remains baffling – indeed, a scientific study set up by the American government called it 'a challenge to the analyst'. The pictures were taken by James McDivitt aboard Gemini 4 when passing over Hawaii on 4 June 1965. They supposedly show a UFO shaped like a 'beer can' but the still released from the film show two fuzzy lights. It appears that both NASA and McDivitt were immediately aware this was not the real object McDivitt photographed, which could not be found during his search of the huge number of pictures taken on the mission. Attempts by sceptics to persuade him that what he saw was merely his own booster rocket flying in orbit have not been successful.

McDivitt still claims to be puzzled by his experience. For a full report see *The UFO Conspiracy* by Jenny Randles.

Another problem comes when dealing with conversations between astronauts and the ground. On several occasions the astronauts did report what might seem to be UFOs, for instance: on Gemini 7, Borman and Lovell reported

'We have bogey at 10 o'clock high.'
'Is that the booster or is that an actual sighting?'
'We have several, looks like debris up here . . . Actual sighting.'

Although nobody has positively identified this debris (seen as tiny lights) it seems a wild leap of logic to deduce that they were alien craft. Tim Good quotes a letter from Lovell (via a third party) in which he insists he saw nothing he could not explain while in space. Does this presumably mean that the Gemini 7 sighting was resolved as something mundane? Of course, some people will doubtless continue to believe it means that he is covering up.

In later missions, peculiar words were spoken when odd things happened. The Apollo 11 crew – after weird but explainable sound effects came over the radio – were asked by NASA: 'You sure you don't have anybody else in there with you?' Apollo 12 saw what they presumed to be panels from the lunar module tumbling in space and as they

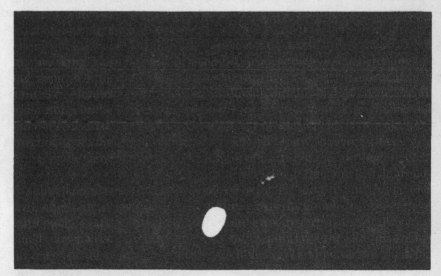

The photograph released by NASA which supposedly shows an unknown object filmed by astronaut James McDivitt while in orbit above Hawaii, on board the Gemini 4 spacecraft. McDivitt later disputed whether this is what he saw and filmed. The debate still rages. (*NASA*)

debated how one seemed to move away at great speed, control said: 'Well, assume it's friendly anyway, OK?'

We do not have the least problem accepting that these were what NASA insist they were – jokes! After all, astronauts are human and were in a very stressful situation, where humour can be a great asset to relieve tension.

Of course, some will continue to seek stranger explanations for what was seen and said. However, what puzzles us is the double standard here in evidence. It is claimed that the reason we do not have definitive proof of alien life from these missions is that the communications were censored before being relayed to the world, so any probative statements about other intelligences were cut out or hidden under the guise of 'technical problems' which sometimes caused transmissions to be lost. On the other hand, the above examples of 'alien contact' and suspect photographic proof are supposed to have somehow escaped this incredible security net.

Yet, surely, if there were any real contacts they would have been obscured, which means that the cases we know about are meaningless. And, if there were no contacts at all, would we not have precisely what we do have – rumours from believers but constant denials from NASA? This absence of evidence is only perceived as proof of a cover-up by those who seem to want such proof. Nobody else is at all satisfied by the claims.

It is easy to take the words of astronauts the wrong way. For instance, the most outspoken of all is undoubtedly Gordon Cooper who flew in Mercury 7. Now working in the Walt Disney Corporation, he has a deep interest and apparent belief in UFOs and actually addressed a debate at the United Nations when this subject had a hearing in 1978.

Cooper saw some strange lights in his pre-NASA days which hardly sound like spaceships from another planet. He referred to swarms of round bright lights that passed over a German air base in waves over several days. Some sort of natural solution seems to be likely. For example, in some similar circumstances we have found that UFO formations similar to these turned out to be high-flying birds reflecting street lights.

Nonetheless, Cooper's views are clearly honest ones forged by his personal experience, but not while in space. He did tell the UN that astronauts are usually reluctant to debate UFOs, because of 'the great numbers of people who have indiscriminately sold fake stories and fraudulent documents abusing their names and reputations'.

However, Tim Good also quotes from that speech where Cooper says, 'There are several of us who do believe in UFOs and have had occasion to see a UFO on the ground or from an airplane.' This provoked speculation about astronauts (even Cooper himself) seeing a *landed* UFO – which would indeed be remarkable. Yet from the context of the quote, it seems just as likely to us that Cooper was really noting how some astronauts had seen UFOs while *they* were 'on the ground' or in 'an airplane'.

This problem of terminology dogs the most famous of all unsupported stories – that the Apollo flights were terminated because NASA found aliens on the moon. Of course, this has to be set alongside other claims that NASA never even went to the moon in the first place and that the whole thing was a stage show set up by special-effects experts! This was an idea later reproduced in the excellent science-fiction movie *Capricorn One*.

The story of how 'NASA met aliens on the moon' is one that seems destined to live on. Occasionally, we still hear of new versions of it. Indeed it has grown with each re-telling and the latest ones are quite colourful.

The original sources seem respectable – former NASA employees Otto Binder and Maurice Chatelain (ex-chief of the communications systems). Yet, except for UFO buffs, few people regard the stories with any seriousness, pointing to works such as Chatelain's 1980 book *Our Ancestors Came From Outer Space*, for support.

Chatelain also claims that the code-word 'Santa Claus' was used for alien life, citing the famous remark (which millions of people heard) when Apollo 8 became the first piloted spaceship to fly around the moon as a prelude to the later landings. The mission had been carefully timed so that this wonderful moment in space flight would take place on Christmas Day. The comment by astronaut Lovell, 'Please be informed that there is a Santa Claus', seems far more likely to have been a carefully planned wisecrack to mark the historic occasion, rather like Armstrong's words when he stepped out on to the moon's surface some months later. Or are we to assume that his 'small step for man' and 'giant leap for mankind' are also codes for something else more ominous?

Some people think they were. Here is their version of events. Both Aldrin and Armstrong saw some UFOs in a crater and Aldrin filmed them as Armstrong got out. The (censored) words spoken are supposed to have been: 'These babies are huge, sir . . . enormous . . . I'm telling you there are other spacecraft out there . . .'

Tim Good quotes an alleged conversation of an unnamed friend of his from 'military intelligence' who reputedly spoke with Neil Armstrong once about this story. Armstrong is supposed to have confirmed that 'We were warned off. There was never any question then of a space station or a moon city.' Yet both of these do figure in NASA's plans for the future.

The launch pad at the Kennedy Space Center at Cape Canavaral is prepared for take-off at the start of another successful Shuttle mission, in October 1989. This was just seven months after *Challenger* was allegedly buzzed by an alien spacecraft while in orbit! (*Jenny Randles*)

Of course, we can do no more than record these tales and comment on the difficulty we face in accepting them. Jenny has spoken with one radio ham (now a top record producer) who even claims that he heard the conversations from the moon and that Armstrong and his fellow astronaut appear to have seen aliens too. Ultimately, all of this comes down to uncorroborated personal testimony.

The most recent encounter allegedly involving NASA astronauts in a head-to-head with extraterrestrials occurred during one of the first missions of the Space Shuttle following the *Challenger* tragedy, when a Shuttle exploded on take off.

The mission was coded STS-29, with a short, several-orbit run in mid-March 1989. The story of its reputed close encounter broke two weeks later in a British newspaper, the *Daily Star*, on 30 March. No other newspaper or major information source in America carried it, save the *Weekly World News*, citing the *Star*. This American paper has a sensational style of reporting which stretches most people's imagination.

Being interested in UFOs, the *Star* had chosen to devote the front-page banner headline to the tale, with SHUTTLE CREW SAW ALIENS. As it turns out, this was misleading even on the strength of the actual claims. The most the crew are supposed to have seen was an alien spaceship, not aliens themselves! The story appears to have come direct from ufologists.

By coincidence, Jenny was at the Granada TV news studios in Liverpool very early on that day, ready to do a live feature on UFOs for the popular networked show *This Morning*. Presenter Richard Madeley soon brushed away the dawn cobwebs by laying down this amazing front page. A debate ensued as to how much we should refer to the matter on air.

Essentially, the claim by the *Star* was this. Reputedly, the Shuttle crew had seen an 'alien craft' which locked itself on to the controls of the billion-dollar piece of NASA hardware, creating a power loss. This information came via radio hams who had heard the censored broadcast. Another passage featured the words: 'Houston, we have a problem. We have a fire.'

For some reason, the *Star* stated that it was believed that 'fire' was a secret code-word NASA employed for UFO (Santa Claus having now been exposed, one presumes.) As perennial ufological critic Andy Roberts rightly commented soon afterwards: 'This does rather beg the question of what astronauts say if they really do have a fire . . . would it be "Houston the shuttle is on UFO"?'

Despite NASA denials, the paper added that British and American UFO experts were 'convinced', which was odd, as this was the first British ufologists had heard of the matter!

The exception appeared to be our friend Tim Good, who is one of the few lucid and persuasive defenders of the extraterrestrial hypothesis in an otherwise conservative British UFO movement. Erroneously called 'Tim Gold' by the *Star*, he suggested that NASA may have to reveal the truth thanks to the American Freedom of Information Act. It is unclear whether Good was the paper's actual source of the story, but he does appear to have been a primary information channel.

The origin of the *Star* story is explained in the *MUFON Journal* in April 1989. This told how one of their investigators, Donald Ratsch from Maryland, monitored Shuttle messages via the Goddard Amateur Radio Club. At 06.35 (local time) on 14 March, he heard the words, 'We have a problem – we have a fire', which *Mufon* speculated to be just that – a fire – and a precursor to some short-lived electrical faults that are known to have occurred on board the Shuttle flight.

However, seven minutes later came the key statement from which all the furore developed. Ratsch reported it as: 'Houston – *Discovery* . . . we still have the alien spacecraft under observance.'

This was tape-recorded by him and the tape does indeed clearly contain these words. It was not certain whether the speaker was supposedly the voice of mission commander Michael Coats or pilot John Blaha. However, to the ear, the voice of John Blaha and the voice on the alleged message from the Space Shuttle do sound remarkably similar. Blaha's name cropped up in the *Star*, but there was no confirmation that he was allegedly responsible.

As the *MUFON Journal* reported in July 1989, the tape would be 'of very significant importance, if true. This was potential evidence that our astronauts may have sighted and confirmed the presence of alien spacecraft over an extended period of time during this flight.'

MUFON researcher Larry Bryant, utilized the Freedom of Information Act to appeal for data from NASA. He requested a complete recording of the ground to space transmission, an official conversation transcript, all photographs taken by the astronauts of the 'alien spacecraft', all radar recordings of the flight and the intruder and copies of all intercepted transmissions by the alien spaceship. This list seems almost to presume that the reported incident had probably taken place and that evidence for this was potentially available. It also was accompanied by an expectation that NASA could or would provide MUFON with conclusive proof.

NASA's response was less conclusive. Bunda Dean of the public relations department in Houston replied on 12 April with the news that anyone could buy a complete audio copy of the flight transmissions for several hundred dollars. Because of this, transcripts of the conversations were not made.

As for the other requests, since 'no such recordings and no such photographs exist', it was evidently not possible to provide them. Dean added, 'We believe that this is a fictitious event and is a hoax perpetrated by a rogue radio operator or an unlicensed person using radio equipment and broadcasting on a repeater frequency that some ham groups use to relay NASA transmissions.'

MUFON president Walt Andrus pondered whether there might be some element of 'official party line' to this reply and noted how John Blaha appeared 'belligerent' when asked about the transmission on a live radio interview, after otherwise calmly discussing the rest of the mission. This may have been true, or Blaha may have just been fed up of all the questions.

MUFON attempted an interesting experiment, using voice analysis, to try to prove that the person speaking the words on the tape was one of the Shuttle crew. This required comparison with words the crew had definitely spoken. Research officer Bob Oechsler (who interviewed Blaha on radio) worked with others to get this analysis done. Awaiting the results, Andrus noted that if positive, 'plans were already formulated between the Fund for UFO Research and MUFON to issue a press release announcing the incredible news.'

However, the results were not positive. There were not sufficient words in the alleged message to allow for anything better than a guess. Concerning the words in the message the MUFON report said it had reached a 'no decision' verdict and found 'no similarities'. MUFON sensibly concluded that the matter should be quietly abandoned and that 'if new evidence is forthcoming on the Discovery tape, the analysis will be resumed and expanded.'

Meanwhile, back in England the Independent UFO Network was doing its best to untangle the truth. Philip Mantle, also a BUFORA director and MUFON member, wrote to NASA and both Blaha and Coats. He got quick and astonishingly helpful replies from them all.

NASA's Michael Braukas said on 3 May (letter reproduced in *UFO Brigantia*, summer 1989) that the events 'are a result of what apparently was a hoax.' The idea of a rogue radio operator was repeated, but then justified by saying it was a 'simple deduction to make since the comments attributed to the STS-29 astronauts were not heard over the space shuttle's communication down link by NASA personnel on duty here at Goddard and the Johnson Space Center, Houston, Texas.'

Colonel John Blaha, belligerent or not, said categorically that 'These statements were not made during my Discovery flight in March 1989.'

Mission commander Michael Coats was even more willing to be frank. In his response, dated 30 May 1989, he noted that 'You cannot believe everything you read in magazines or newspapers. The reason you will never hear an actual tape recording of any of us on Discovery discussing "aliens" is because we never did. The stories are amusing, but pure fiction.' Coats commented on the supposed code-word that was used. He said, firmly, ' "Fire" means fire, which is much more alarming than aliens ever would be.' Given the closed confines and impossibility of any rescue in the loneliness of space, one can well imagine these words to be true. Coats concluded, 'If we did see any aliens, the whole world would hear about it immediately. We are just as curious about the possibility of other life as anyone, so why would we try to be secretive?'

We have concentrated on this case because it offers a very important opportunity to discuss the controversy surrounding UFO sightings in space. With many of the supposed astronaut sightings, it is impossible to make definite statements because they happened so long ago and they are now an established part of space-age folklore. The *Discovery* Shuttle encounter may eventually end up that way, even though it is fundamentally flawed.

Why would the mission continue and run its planned course if the Shuttle had been damaged by a power black-out after a close encounter with a UFO? Furthermore, if 'fire' is a code-name for UFO, surely there would be also be a code for 'alien spacecraft'. This is far too presumptive a term for the military-dominated space programme to use. Is it not more likely that words such as 'unknown target' would be spoken in such a situation?

In addition, nobody seems to have wondered whether 'alien spacecraft' might not have referred to a mundane device from another power on earth, such as a Soviet spy satellite which was being monitored in understandable secrecy.

Fortunately, some sensible and speedy investigations by much-maligned ufologists have taken place on both sides of the Atlantic and quickly put the rumours of a Space Shuttle encounter to rest. That will probably not prevent it from turning into yet another supernatural legend and appearing in books of tall stories throughout the next century.

But those of us interested in the true and scientific search for alien life forms must dismiss this account, perhaps along with most other alleged encounters between extraterrestrials and our astronauts.

18
Crashed Spaceships and Dead Aliens

Did an alien spacecraft explode over Russia in the early morning of 30 June 1908? That was the astonishing premise explored by John Baxter and Thomas Atkins in their book *The Fire Came By*. Even without such speculations, the facts themselves are riveting.

At about 7 am, a fiery object, which seemed brighter than the sun, entered the earth's atmosphere and came down over the Tungus Forest in Siberia. As it appeared over the Taiga River, it became transformed into a fiery column reaching up into the atmosphere. A devastating series of thunderclaps came from the phenomenon, heard over 60 miles (90 kilometres) away.

After the thunderclaps, a shock wave followed which tore off roofs and uprooted trees up to 70 miles (110 kilometres) distant. Earth tremors were registered by seismological stations in Germany and Russia. A weaker blast which was registered in Britain, occurred after the first explosions, and the shock wave travelled twice around the earth. Three days later, luminescent clouds were observed over Europe and North Africa, bright enough to allow newspapers to be read and photographs to be taken at night. The light was described as 'pink' and 'yellowish-green.'

Baxter and Atkins and many other commentators have noted the many similarities between the Tunguska (the local name of the small area where the event occurred) disaster and the detonation of nuclear weapons. As no nation had that kind of destructive force at the turn of the century, they conjecture that the explosion was caused by a malfunction of a nuclear engine aboard an alien spacecraft.

Because of its remoteness, it was 20 years before a scientific expedition went to the area, led by astronomer Leonid Kulik. Kulik was convinced the explosion must have been caused by a giant meteorite impacting with the earth, but no crater was ever found. Quite the contrary, at the impact site, trees were standing, although branches had been ripped off by a downward blast. This was inconsistent with an explosion on the ground such as a meteor impact would produce.

But more than that, no large meteorite samples were found at the site, although later expeditions did find radiation. It was speculated that this might be due to fallout from the USSR's atmospheric atomic tests during the 1950s and early 1960s.

Aerodynamics expert A. Y. Monoskov was also against the meteorite theory. He analysed the testimonies of a large number of eyewitnesses to the Tunguska event and concluded that the object, prior to exploding, had slowed down. It was also shown to have changed course twice.

Ufologist Kevin Randle interviewed physicist Dr James A. Van Allen in 1977 about the incident and the ever-popular, if unestablished, possibility it could have been caused by an interstellar craft. Dr Van Allen is, of course, best remembered for his discovery of the 'Van Allen Belt' – bands of geomagnetically trapped radiation encircling the earth – and he was also a friend of ufologist, the late Dr Allen Hynek.

Van Allen agreed with Russian meteor expert E. L. Krinov, who had compiled masses of data regarding the possibility that the cause had been a comet. Comets are comprised of a relatively small solid nucleus surrounded by a much larger region of volatile gases and dust. The substantial amounts of small magnitites and silicate globules found at the impact site would be characteristic of meteorite falls, he said.

Randle cited an investigation in 1959 which discovered traces of radioactive Caesium 137, but because of atom-bomb tests in the area, Van Allen concluded that the caesium in itself may not be significant, although a more intense investigation into the distribution and radioactive level of the material would decide the matter. However, he agreed that a comet cannot change direction; neither could a space vehicle which had entered the atmosphere in trouble, he added.

The belief that a comet, and not a spacecraft, was the source of this terrific explosion can be viewed as little more than tentative; although many scientists are intrigued by the suggestion. With so many objects (planets, asteroids, meteors, comets, etc.) whirling about in continual orbits, the odds of this celestial snooker game do imply that sooner or later one of these objects is bound to hit the Earth with devastating consequences, rather than continue to pass narrowly by as comets do several times each century.

However, neither before, nor since that day, have we experienced anything quite like the Tunguska explosion – not at least in historical memory. There are those who believe that whatever caused the destruction in the tundra may also have wiped out the dinosaurs when a similar thing happened 60 million years earlier on a much more dramatic scale.

So what *was* it? Atomic explosions do not occur in nature. And this thing, according to trigonometry, changed course, not just once, but twice. Perhaps it is a quantum leap to speculate on the spaceship idea, especially given the lack of hard evidence. Perhaps it is just coincidence that the explosion mirrored modern atomic blasts.

The desert around Roswell, New Mexico. This photograph was taken in July 1987, 40 years to the day after an alien spacecraft supposedly crashed there and was captured by the American government. *(Jenny Randles)*

Scientists will continue to probe for the answer to what was no mean explosion. It was more powerful than the combined forces of the bombs which utterly devastated the cities of Hiroshima and Nagasaki 37 years later.

Tunguska is not the only place where an out-of-control spaceship has allegedly visited earth, according to some commentators at least. There has long been a rumour in the UFO field that an alien spacecraft crashed into the deserts of New Mexico in July 1947. The US Air Force supposedly gathered up the wreckage, together with the badly decomposing bodies of six 4-foot (1.2-metre) tall aliens, which were preserved in ice, and the whole affair was hushed up.

What is the truth behind this and other incredible stories that piloted extraterrestrial craft have occasionally malfunctioned and crashed to earth? Why are they always in such conveniently remote locations, such as the western deserts of America (a favourite haunt)? And is it in any sense relevant that these same regions are where atomic weapons and space technology were first developed by trial and error, often in understandable top secrecy?

The New Mexico crash became known as the Roswell incident – Roswell being the name of the nearby town, as well as of the base from which the wreckage was supposedly flown to its final destination at Wright-Patterson Air Force Base in Ohio. No one who has critically studied the evidence is disputing that something crashed. The question is: *what*?

In the late evening of 2 July 1947, a disc-shaped object was reputedly observed passing low over Roswell, heading in a northwesterly direction. Later, in the desert to the northwest, over a remote region of ranch land, during a violent thunderstorm, a striking and unusual explosion was heard and noted by local people and military personnel at Roswell Air Base. The following day, ranch manager 'Mac' Brazel, his daughter Bessie and son Bill, discovered some strange wreckage on their land. But as they were not on the telephone, it was not reported until several days later, when Brazel went into Roswell.

There, on 7 July, Brazel informed the authorities of a huge sheet of metallic-like material which he had dragged, using a pick-up truck, from the site and into an outbuilding. One of the first men on the scene was journalist Johnny McBoyle. After examining the wreckage, he returned to the radio station that he part-owned and phoned through to a sister-station, KOAT in Albuquerque. The reason for this was to use their teletype machine to inform the world. This is what KOAT employee Lydia Sleppy reported he said:

'Lydia, get ready for a scoop! We want to try to get this on the ABC wire right away. Listen to this! A flying-saucer has crashed . . . No I'm not joking. It crashed near Roswell. I've been there and seen it. It's like a big crumpled dishpan. The whole area is now closed off. And get this – they're saying something about little men being on board.'

As Lydia started to type the report into the machine, it stopped after a few sentences. She informed McBoyle, but he seemed to be arguing with someone else in the background. When he did come back, he seemed tense and under pressure, and although he promised to talk to her later, she never heard from him again. Just then, the teletype started up again, but this time it was addressing her: ATTENTION ALBUQUERQUE: CEASE TRANSMISSION. REPEAT. CEASE TRANSMISSION. NATIONAL SECURITY ITEM. DO NOT TRANSMIT. STAND BY.

Another station in Roswell, KGFL, also tried to transmit a story, but was stopped by a long-distance call from the Secretary of the Federal Communications Commission in Washington, DC, and from New Mexico's Senator Chavez. According to the owner's son, Walt Whitmore Jr, both these men threatened to revoke the station's licence if his father did not comply.

The preliminary investigation of the crash was carried out by Major Jesse Marcel, staff intelligence officer for the Army Air Force at Roswell Air Base, and Counter-Intelligence Corps officer 'Cav' Cavitt. At that time, Roswell was home of the 509th Bomb Group of the US Army Air Force – the only combat-trained atom bomb group in the world.

When the officers returned with the debris, Lieutenant Walter Haut, base public information officer, issued a press release on the direct orders of base commander Colonel William Blanchard:

The many rumours regarding the flying disc became a reality yesterday when the intelligence office of the 509th Bomb Group of the Eighth Air Force, Roswell Army Air Field, was fortunate enough to gain possession of a disc through the cooperation of one of the local ranchers and the sheriff's office of Chaves County.

The flying object landed on a ranch near Roswell some time last week. Not having phone facilities, the rancher stored the disc until such time as he was able to contact the sheriff's office, who in turn notified Major Jesse A. Marcel of the 509th Bomb Group Intelligence Office.

Action was immediately taken and the disc was picked up at the rancher's home. It was inspected at the Roswell Army Air Field and subsequently loaned by Major Marcel to higher headquarters.

Was the press statement a mistake? If so, what kind of mistake? Certainly things moved swiftly to defuse the furore it created once it was out. The release was enthusiastically picked up by the media and the story eventually circulated about the world. But the army effectively and quickly deflated the crashed UFO interpretation of the incident at the skilled hands of General Roger M. Ramey.

Marcel intended flying with the debris on a B-29 all the way to Wright-Patterson, but the aircraft stopped off on the way at Fort Worth (now Carswell Air Force Base), Texas. It was there that General Ramey took charge of the situation, allegedly forbade anyone from talking to reporters, sent Major Marcel back to Roswell, and then issued a press statement which some say contradicted the observations of his own men, reporter Johnny McBoyle and farmer Mac Brazel.

According to Ramey, the debris was nothing more than the remains of a weather balloon with a tin-foil radar target attached to it. He even produced one to prove it, but dedicated civilian investigators now believe the real wreckage was by then safely on its way to Wright-Patterson.

There were indeed a number of balloon projects in progress in the area at that time, but although many weather balloons were mistaken for UFOs in flight, especially in those early days when UFOs were very much a new media issue in America, it is hard to imagine a grounded object being so badly mistaken. In fact, during that period, two balloons did crash and they were recovered without any fuss. Balloon expert Irving Newton, brought into the affair by General Ramey to verify the alleged debris, commented that the officers at Roswell would surely have recognized it for what it was. Recently, ufologist John Keel suggested that the balloon was Japanese. If so, secrecy and confusion at the time would be more understandable.

If it was just a weather balloon, why was rancher Brazel reputedly held incommunicado for almost a week by the authorities – afterwards changing his story to the media? The accumulating evidence strongly implies that the weather-balloon story was a hastily concocted cover and that whatever really occurred had military security implications.

It would seem, too, that the debris was scattered over a wide area, that perhaps there had even been two crashes – two objects which had collided, or one object which had shed some of its structure at Roswell, but had managed to continue in flight for 150 miles (240 kilometres) to come to rest near Magdalena. Allegedly, it was in this second part of the wreckage, discovered by soil conservation engineer Barney Barnett and some students involved in an archaeological dig, that the tiny bodies of aliens were found. However, this testimony is much less documented than the rest.

Despite Barnett's boss, J. F. Danley, recalling hearing about the incident at the time, no corroborative evidence has been found, or additional witnesses so far traced. This means that, unlike the Roswell incident, which offers strong evidence for the crash of some kind of craft, the tales about dead alien pilots at Magdalena are far from established. But even if they are unverified (which most careful scientists will almost inevitably conclude), this need not affect the status of the primary crash at Roswell, where something does appear to have happened that led to near panic in the military hierarchy.

Most of the research into this case has been carried out by ufologist William Moore and his colleague, nuclear physicist Stanton Friedman. In 1980 they published *The Roswell Incident*, with Charles Berlitz. They have since periodically updated their work via MUFON.

However, a major reinvestigation took place in 1989 when the Center for UFO Studies mounted a five-day expedition to the remote and rattle-snake-infested crash site to carry out much new work. Mimi Hynek, widow of the famous astronomer, was present along with eight others, including two ufologists (ex-military intelligence man Kevin Randle and the Center's director of special investigations, Don Schmitt).

Randle and Schmitt have published a book on the discoveries. The first reports on some of these appear in an article by the Center's scientific director, Mark Rodeghier ('Roswell, 1989', *International UFO Reporter*, volume 14, number 5), who was part of the expedition. The new study interviewed 'over 150 individuals who are in some way connected with the crash' (many had never been interrogated fully before) and also did something else not previously reported: conducted a detailed scientific analysis of the precise landing site for any hint of crash debris still remaining. None was found.

In the course of these detailed investigations, ufologists have invested a small fortune and many hundreds of hours chasing up government documents and witnesses. Their determination may now be paying off.

Major Marcel said that when he got to the site there was a lot of wreckage scattered over about three-quarters of a mile (1 kilometre), but no complete machine. Marcel was adamant that it was not the remains of any kind of weather balloon or missile, as he was very familar with both. The debris consisted of some very lightweight 'beams' of an unknown material, which had weird hieroglyphics on them, a small black metallic-like box, a great deal of a brown parchment-like material, which was extremely strong, and a large number of pieces of 'tin foil', except it was not tin foil.

The material would not burn, and the 'tin foil', although paper thin, could not be permanently bent, even with a sledge-hammer. Major Marcel said it was like a metal, but with plastic properties. The major thought that the object had blown up in the atmosphere. All of this was backed up by his son, who is now a doctor.

Bill Brazel, son of the rancher who first found the wreckage, assisted in the 1989 investigation and remembered that after the crash he recalls seeing a shallow gouge in the earth as if the object had skidded and skipped up into the air and come down again.

Bessie Brazel, the rancher's daughter, also verified the strange 'numbers' and 'letters' arranged in columns on the long narrow beams. She also remembered some tubes like pipe sleeves.

Brigadier General T. J. Dubose, who was adjutant to General Ramey at the time, said that the weather-balloon scenario was a complete fabrication, foisted on Ramey from elsewhere.

For a run-of-the-mill balloon crash, it seems odd that the military should reportedly erect road-blocks and spend several days scouring the site for every tiny bit of debris.

Are all these witnessess telling the truth? That most are, seems a sensible option. To imagine that they are all part of a disinformation campaign seems improbable. However, even accepting the witnesses at their word, all it tells us is that some kind of strange craft apparently crashed. Could a secret American experimental aircraft or rocket have been the culprit? Perhaps the technology was so experimental that nobody involved in the military site investigation knew about it. They may even have presumed it was a spaceship.

This possibility has been mooted by a few researchers who have studied the evidence, such as Ron Schaffner. In his article 'Roswell: A Federal Case?' (*UFO Brigantia*, summer 1989), he offers plausible arguments that the story may relate to a malfunctioning rocket. There may even be a possibility that the stories about alien bodies refer to monkeys who were the unfortunate victims of some prototype experiments which went wrong. Given the Cold War and the race to develop intercontinental ballistic missiles, there is little doubt that if any of this were correct, most of the staff at Roswell Air Base would have been unaware of what was going on.

One thing seems clear: whatever crashed was not an American balloon, but that is hardly prime evidence for an alien spaceship either.

There have been other rumours of UFO crashes which allegedly have been hushed up by the authorities. Most, if not all, are considered at best unproven, and some, very probable hoaxes, even within ufology. They continue right up to the present day, with a 1989 story of a crash in South Africa where the spaceship was supposedly shot down by a sophisticated laser canon. This remains unsubstantiated and will probably stay that way. Few people regard any of the latest stories as verified to anything like a scientifically acceptable standard.

In recent years, documents reputedly 'leaked' from covert intelligence sources have found their way into the UFO community. They allege that a secret board of scientists and politicians, known as the 'Majestic 12', was formed in America in the wake of the Roswell crash. Their job was to act as a think-tank on the UFO phenomenon, reporting directly to the president. There is conflicting evidence on authenticity of these documents, but the majority view within ufology seems to be cautionary. Magazines, like *New Scientist*, have outspokenly rejected them. The people listed as Majestic-12 members at the time of the 'leaked' file (that is, 1952) are now dead, so little can be done to establish the credibility of the claims, although some investigators have tried valiantly.

Along with these documents, rumours emerged of supposed autopsies carried out on the alien crew members. Although these rumours originate in America, we personally became embroiled within a mini-James Bond scenario in Britain which may offer clues about the overall picture.

On 28 October 1986, Jenny received a phone call from a man who refused to give his name or phone number. He said his commanding officer had supplied her number because he wanted certain information to be handed on. The stranger gave the impression he was just the messenger-boy. He referred to six reports in his possession, spanning 600 pages, and names like Wright-Patterson Air Force Base and Dr Frederick Hawser (or Hauser) were mentioned. Many other codings were cited which matched what was later revealed in the Majestic-12 papers. The files we were offered supposedly contained data from a scientist conducting biological analysis on alien bodies recovered from flying-saucer crashes.

Our source said he believed one report was dated 1948. There were other references indicating that the Roswell crash site was being discussed. Another file was reputedly from 2 October 1977 and bore the sinister title: ELIMINATION OF NON-MILITARY SOURCES.

At first we expected this to be a single hoax phone call, but the man called back on 30 October, saying he was now willing to give his name and home town in Lancashire. He wanted to see us, so we arranged a meeting in a pub near Manchester. We gave our names at the bar, but no one had asked for us. We were on the point of leaving when a young man in his late twenties came forward and sheepishly introduced himself.

During the next few hours he gave us a long and extremely detailed account of how he came into possession of the alleged files. He related in depth their contents and gave extensive replies to all our questions, and accepted with apparent cool our outspoken scepticism and insistence we would have to 'check this out'.

The man, whom we shall call John, had been in a branch of the British Army until February 1985. His commanding officer had spent time on attachment in America, where he had become friends with a US Air Force officer at Wright-Patterson. This officer was a computer technician who claimed that, while investigating a fault, he accidently tapped into some secret UFO files. He took copies and was then arrested for being in a secure area without permission. Despite interrogation, he did not admit to having any files, but managed to inform his British friend where they were hidden, and implored him to get them out of the country. He felt they were much too important to be kept from the public. A few days later, while on remand awaiting a tribunal, the American officer died in a car crash – officially while drunk. This supposedly occurred in 1983. John's commanding officer believed that the man was murdered in a conspiracy.

According to John, once the files were in England, his commanding officer set about testing all of his men over a two-year period to discover which of them held the same views as he did on the UFO subject. He was evidently seeking a trusted confederate to help with the dissemination of the files. As the process evolved, John was shown confidential UFO reports and a daylight photograph of an unknown craft to test his response.

Finally, John left the army, but in 1986, when he returned for a weekend training camp for reservists, his commanding officer told him the full story of the files.

John was given the key to a flat and told to pick them up there, study them, then hand them over to Jenny Randles if he judged that this was the right thing to do. Knowing little about the UFO subject, the files made no sense to him, so he initially only scanned through them. The first time he had read them fully was the day before our meeting, realizing, at last, the explosive nature of their content. It was only then that he feared for the safety of his wife and three children. To protect himself, he split the documents up and hid them, waiting to see what we were like before daring to hand them over.

We told him that his story was interesting, but he had not presented us with a single item of material evidence and he must understand that we could not accept such an incredible yarn at face value. He agreed to meet us eight days later at a country park, by then he would have had time to collate the files and would present them to us as an example of good faith. We arrived in plenty of time at the rendezvous, but John never showed up.

Eleven days later after this aborted meeting, when we had more or less concluded that the whole episode had been a hoax, Jenny was surprised to receive a letter from John, explaining what had supposedly gone wrong. In it, he claimed that two days after our discussion in the pub, 'I was invited to my home base to assist in an internal investigation.' This, he said, involved force, and meant he had to remain on the base for several days while the matter was completed. He was interrogated about 'sensitive' documents and told that these were 'the creation of an educated prankster' to which 'no credence could be attributed.' On release, he was warned it was in his interest that no mention of the documents be made. Apparently the documents had been traced by the military, and John was now embarrassed that he could offer us no proof, and hoped we would not think he was a hoaxer but understood completely if this was our view. He felt he had spread enough information to make sure that his own safety was secure and promised one day he would vindicate his story.

On the face of it, this seems a ridiculous tale, and of course it is virtually impossible to check out. We did try, even before our appointment to collect the files, and that may have indicated our basic scepticism to our informant. Or he may have just got cold feet after seeing our apparent caution.

We were able to note the make of his car and its licence number. This demonstrated that the vehicle was registered in the precise month when he claimed to have left the army. We were not convinced that John was just a hoaxer and felt it more likely he had been used, just as we had, possibly to leak disinformation.

It did not seem like a coincidence to us that three groups of investigators were all approached and, unknown to each other, offered secret files on UFO crashes and the alleged team of scientists officially set up to study them. This happened more or less at the same time! The offer to us came just as Jenny was in the very last stages of a book on the alleged UFO cover-up, published later as *The UFO Conspiracy*. She chose not to relate this uncorroborated tale.

Timothy Good broke the news of the so-called Majestic-12 documents to the world seven months later, in the publicity run-up to his best-selling book *Above Top Secret*. He says that the papers were given to him by an intelligence source who he refuses to name and that this was in January 1987 (only weeks after our involvement with John). At that time he was also known to be in the final stages of preparing his own book about the cover-up, and he chose to use the story, but he did not have time to check the authenticity of the files.

This all seems very curious and may well suggest that a disinformation campaign was mounted against ufologists who were in the most likely positions of influence at the time. If the files were not genuine and could never be established as such, presenting them to the public via UFO books and other outlets would detract from whatever genuine evidence might be coming out, for example, suspicions about the real (secret) truth behind the Roswell crash.

The full story of the Majestic-12 files has not been told here. Other 'leaks' took place, but this is not the place to debate the issue, although subsequent revelations regarding these files seem to us to be consistent with the view that they were planted into the UFO community from outside for motives which can only be guessed at.

Perhaps somebody knew how adept the UFO field tends to be at shooting itself in the foot, thanks to its understandable but unfortunate tendency to embrace and then eagerly promote whatever evidence comes along. If you sell the customer fool's gold alongside a few scraps of the genuine article, then nobody can be blamed if they don't take the salesman very seriously.

As for the crashes at Roswell and elsewhere, we are left feeling certain that something took place, but a good deal less certain that it has anything at all to do with our search for aliens. But if all this leaked information on crashed discs is disinformation purposely seeded into the UFO community, we still do not know why. Why should governments take such an interest in tinkering with our perceptions of a phenomenon which they have always denied exists?

19
Aliens in Focus

On 7 July 1948 one of the most gruesome photographs ever taken first saw the light of day. It shows the charred and blackened body of what purports to be an alien who died when his spaceship crashed just south of Laredo, New Mexico, a little way over the border into Mexico itself.

As usual with such stories, there is quite a tale behind it. Apparently, the American authorities were tracking the spaceship on radar when it hit trouble. (Alien craft have a tendency to do this as soon as they overfly the south-western American states!) The spaceship hit the ground. There was a fire but substantial remains of a 90-foot (27-metre) wide 'saucer' was allegedly visible to the military rescue team. The area was cordoned off and officers photographed the body of the single occupant. Then the remains were taken away over the border to America, and the site cleared, with debris transported to a secret Mexican base. All this occurred without public knowledge – a common but very clever trick that is rarely explained.

This story is courtesy of the usual, mostly anonymous, sources. It sounds absurd, but it has a twist that other crashed-saucer tales do not possess – a photograph of the dead alien!

According to 'testimony', here is what the being looked like: it was 4 feet 6 inches (1.4 metres) tall, with a large head and eye sockets but no visible nose or ears. While there is no doubting the sincerity of the researchers who collated the information, inevitably, we do tend to have major reservations about the picture.

James Oberg, the space expert who has debunked many of the astronaut cases of alien contact, found numerous problems with the tale. For instance, according to him, aircraft supposedly used in the rescue mission simply did not fly until well after 1948. Others contended that the 'alien' was a monkey that had unfortunately met its death in an early rocket test. However, there is another clue in the photograph not to be overlooked. This is the indisputable proof that aliens can be short-sighted too! The frame from a pair of spectacles is clearly visible.

The truth may be that this photograph is really of the remains of a human pilot who died tragically in his blazing aircraft. If this does prove to be the case, it is all the more sickening that it has been used by persons unknown in order to manufacture a hoax.

The so-called Laredo alien does not stand alone. It is merely the first of half-a-dozen cases that have become part of ufolore. But do any of them offer reliable evidence?

The next picture became famous because it was associated with the Roswell incident that we have already discussed in the previous chapter. This was thanks to a sketch based on a poor photocopy that featured in the photographic section of Charles Berlitz and William Moore's book about the case, *The Roswell Incident*. To be fair, they do say that this picture may, or may not, relate to the Roswell case and that it was generally treated cautiously by American officials.

However, the book reports that the photograph was handed over to the FBI by an unnamed informant in May 1950 who claimed that he had bought it for $1! It reportedly originally surfaced in Wiesbaden, West Germany, shortly before this. The fuzzy image supposedly shows a living alien who survived a crash and was being escorted by two military police.

In *The Roswell Incident*, copyright of the photograph was credited to the UFO Information Network, an American civilian group. Later Dennis Pilchis, who was director of that network in Ohio, published a 'release' (see *Search Magazine*, spring 1983) in which he noted that they found the photograph during a Freedom of Information scan of the FBI files when these became available under the new law. Only a 'bad xerox' of the photograph was available in these records (hence their sketch) and the poor reproduction in the Roswell book. He insists that he made it clear, when sending out copies, that the photograph had nothing to do with the Roswell case. The resultant book does not contend that it was related, but it has inevitably helped foster a view among less critical readers that this picture offers photographic proof of a captured alien.

In fact, the truth is quite different. An investigation by German researcher Klaus Webner (published in *PROBE*, September 1981) made the facts known. He traced the photograph to a story in the newspaper *Wiesbadener Tagblatt* which reported in graphic headlines how A GIANT FLYING DISC CRASHED AT THE BLEIDENSTADTER KOPF. CREW MEMBER IS IN PROTECTIVE DETENTION. NO CAUSE FOR PANIC.

The alien was described as having four fingers, a large oval head and huge glaring eyes (typical of the stories since offered through many other rumours and anonymous witnesses). 'Mr X', as the paper called the alien, was taken to a local hotel 'to get used to our air'. The American authorities were said to be tight-lipped. The article concluded with a good-quality reproduction of the actual photograph depicting the two soldiers leading off the alien, complete with breathing apparatus.

However, the most important clue about this story does not come in

Is this an alien captured by military police in West Germany? This photograph has generated a lot of controversy since it surfaced in the early 1950s. Only recently has the real background to it emerged.
(*Wiesbadener Tagblatt*)

this text at all, but at the head of the newspaper, where the date is customarily carried. This date reads: Saturday, 1 April 1950.

To confirm his suspicions that it was indeed an April Fool's spoof, Webner succeeded in interviewing the reporter (Wilhelm Sprunkel) some 31 years after it was conducted. It seems that flying-saucers and life in space were in the news at the time and the journalist thought them silly, so he decided to concoct a trick photo and story. The American authorities were in on the joke and permission came via several sources to use real military officers. The alien was the 5-year-old son of the photographer, whose appearance was altered by artwork.

The hoax was so successful in 1950 that it provoked a lot of response and it was reproduced in an American newspaper for the local troops. It was explained away to readers two days later (under the heading GOOD LET-DOWN), but even so some people refused to believe this exposé. Sprunkel explained how he argued on the phone with one woman who simply refused to accept his admission and who wanted to purchase the copyright.

The case has long been celebrated locally, but prior to the enquiries by Klaus Webner no ufologist had checked the sources, even though the FBI files made clear where (and approximately when) the photograph had emerged.

On 22 April 1981 the *Wiesbadener Tagblatt* carried a new story, headed A TAGBLATT APRIL JOKE IN FILES OF THE FBI.

An even more baffling picture of an alien captured on film. However, the facts about this case have recently come to light.

April Fool jokes about aliens are not uncommon. As recently as 1989, entrepreneur, businessman and transatlantic balloonist Richard Branson brought traffic to a standstill on one of England's busiest highways, the M25 motorway, when a spoof that he cleverly set up went awry.

Branson had a balloon decked out to look like a spaceship and a midget dressed as an alien pilot. Hundreds of startled motorists witnessed the early-morning close encounter and there were fears that an accident might have been caused by the poor visibility. Luckily nothing serious did take place, although many motorists were more interested in what was happening in the sky than ahead of them on the road. One woman fled outside in panic, forgetting she was stark naked and the police rushed to the scene of this historic landing (near Redhill in Surrey) with truncheons raised and typical British reserve at the ready, possibly preparing to arrest the illegally parked spaceship! Goodness knows what would have happened had these been real aliens complete with ultra-powerful weapons. It is to be hoped if they ever do decide to visit England, they plan their mission rather more carefully.

A photograph similar to the Wiesbaden hoax stems from about 1952 and shows two 'security men' in trench coats leading off a tiny figure, some 2 or 3 feet (nearly 1 metre) tall. It is completely naked and weirder looking than the one from Germany. There has been speculation that this is a monkey with all its hair shaved off, although the face looks chillingly human. Until very recently nothing was known about the origin of this dubious image. Now, thanks to the Scandinavian UFO Information team and their publication *UFO-Nyt*, we have some possible answers.

Ole Henningsen reports various enquiries into the so-called 'silver man' photograph. For example, commercial artist Claus Westh-Henrichsen had studied it in great detail and found many problems. For instance, he notes that the hand positions of the 'security men' indicate that they were gripping a rigid object. After carrying out tests, he proposes that they were actually pushing a pram!

Similarly, it appears that by examining the feet of the two security men and the alien (not fully visible on the print), it transpires that the alien would have to be floating above the ground.

For these and a host of other reasons, Westh-Henrichsen is certain that the picture is another hoax, formed from an amalgam of a shot of the two men and the 'pram', with the 'alien' superimposed over it.

Meanwhile Hans-Werner Peiniger from a West German UFO group alleges that it is yet another April Fool's joke and notes that the authors are G. Falscht and R. Logen, which in English is similar to D. Bunker and A. Fraud (in fact, literal translations would be 'forged' and 'make-believe'). So another classic photograph appears to bite the dust!

One of the most intriguing of all photographs was taken in Britain on Sunday, 24 May 1964, by Carlisle fireman Jim Templeton, who was on a quiet marsh bank of the Solway Firth in Cumbria (looking towards Gretna Green and the Scottish border). It was a lovely summer's day and he was taking photographs of his young daughter, Elizabeth. His wife and other daughter were also present.

Nothing odd was seen by the family, but Templeton did notice that all the cows and sheep in the fields were behaving as if an electrical storm were approaching. He knew the area well and they were certainly acting oddly by huddling together. He took his photographs with his Pentacon camera and sent them away to be processed by Kodak as normal. A few days later they were returned with a comment that it was a shame the man in the background had spoilt the best shot. Everyone agreed there had been no man present, but sure enough, he was there on one single colour photograph.

To be accurate, there is a strange image visible behind Elizabeth's head which looks for all the world like a man wearing a large white padded 'spacesuit' and visored helmet. If it is a real person standing on the ground behind the girl, then he is at least 7 feet (2.1 metres) tall and leaning crookedly. If, instead, he is of normal height, as the proportions tend to suggest, then he is evidently floating above the ground, tilted at an angle.

Every attempt to explain the photograph has failed. Kodak later told

British fireman Jim Templeton took this photograph of his young daughter on the Cumbrian marshes. Nobody was seen to intrude into the shot, but the result was a photograph that has astonished the world for 30 years. Is that really a spacesuited figure behind the girl? (*Jim Templeton*)

the Press it was not an emulsion defect or accidental image created by some chance combination of flaws. They even offered a reward of free film for life to anyone who could demonstrate the cause, but the prize remains unclaimed. The Carlisle police, to whom the Templetons took the picture, speculated about an accidental double exposure – someone else's negative being superimposed on to the one taken that day by Templeton – but close inspection shows no trace of the 'spaceman' where the girl's head fills the picture. A double exposure would presumably superimpose on to this part as well as on to the light sky background. In the end Chief Superintendent Tom Oldcorn concluded vaguely, 'It must be some sort of freak picture.'

In this case, the alien nature of the being is merely the product of little more than speculation after the fact. True, it looks like a being in a spacesuit as then worn by NASA and Russian astronauts. However, there is another curious twist. Out across the estuary to the north, in the direction the figure seems to be looking, there was a NATO radar complex and an atomic power station at Chapel Cross. There was also a telecommunications site and a missile-launching range. Few locations in Britain could be more fascinating.

Given these facts, the figure might just as easily be perceived as somebody wearing nuclear protection clothing; although this scarcely makes the resultant image any less baffling. A Japanese research team did suggest that a worker at one of these facilities may have been the source of a freak optical effect, with perhaps a holographic image projected across miles of space due to unusual weather conditions. However, staff at the atomic station point out that nobody was working in radiation suits on that day.

For a time, police sources claimed the case was solved. A local eccentric was known to run naked around the marsh. Was this him? Jim Templeton had seen him, but not on that particular day. To be honest, the semi-transparent look to the 'spaceman', its size and crooked stance on the photograph all combine to make this 'streaker' solution quite ridiculous.

Two or three months after the story was made public, there was another mystery for Jim Templeton to fathom out. It seems that two strange men had phoned to ask for a meeting out on the marsh where the picture was taken. They arrived, dressed in dark suits, driving an expensive Jaguar car, and referred to one another only by numbers! They showed identity cards which implied they were from some branch of the government and quizzed Jim intently about the weather on the day the picture was taken and the behaviour of the local animals – for instance, were any birds around? They seemed intent on getting him to admit that a real man was at the location that day, but he insisted that nobody had been, so they left in a huff. Stranded, Jim had to walk the 5 miles (8 kilometres) home!

Visits by two official-looking men in a large car who then behave oddly and ask peculiar but specific questions are among the most baffling aspects of the UFO phenomenon. In Britain we have logged a dozen or more similar cases and, although usually not publicized, they are amazingly consistent. They seem to be common when anomalous photographs are taken.

The Solway Firth case is challenging. It features widely in magazines but normally without information about its background. We decided to put that right here and are very grateful to Jim Templeton and his wife for answering all our questions very courteously. We were completely satisfied that they were faithfully reporting exactly what took place 26 years ago and, as Jim said: 'We have always hoped that somebody else would take a similar photograph, but as yet nobody has. We are definitely convinced that whoever it is on the photograph is a real person. So far as we are concerned it is a straightforward photograph and he just turned up on it.'

There has been no attempt to make money out of the picture and Jim Templeton revealed that a very famous and extremely wealthy American family stopped by when passing through on a touring holiday of Scotland. He was invited to return to America for a free holiday, but turned this down in typical fashion. So far as the Templeton family are concerned, the event just happened and that was that. It is nothing to make a fuss over.

It appears that other odd things have happened in the same area, especially over the estuary in the area around Annan. There are tales of weird glows, cows not giving milk for no reason, strange burn marks on the ground and a curious light that frightened one driver on the marsh itself, which then winked out into nothingness. But the Templetons have never experienced anything else themselves.

However, just before we spoke to them in May 1990, a life-size print of their 'spaceman' had been made by a new copying process. This was part of a company demonstration for a new type of machine. Jim Templeton said that this huge colour print has added features to the entity that had not been visible before, including ears behind the visor and a sort of breathing tube down the back of the suit.

Clearly this is an amazing story, but do we need to speculate, as ufologist Gordon Creighton did in the November/December 1964 issue of *Flying Saucer Review*? He suggested that the solution to this case might be as follows. A UFO had already landed on the marsh before the photograph was taken. Having rendered itself and its crew invisible, one occupant had accidentally intruded into the picture, but unfortunately for us, the camera did not point at the invisible spaceship!

There have been a few other supposed photographs of aliens, although these tend to have been dismissed from consideration very quickly. One appears to be a crude toy soldier stuck in a lump of Plasticine; another, critics claim, was admitted by the American youth who took it to be a doll in front of a hard-boiled egg illuminated by a flashlight!

There is a curious series of pictures taken by a police officer in Falkville, Alabama, during a UFO wave in October 1973. He was answering a call by a woman who had seen a UFO. His picture shows a figure in a silver suit that reputedly fled, robot-like, when pursued. But as the figure has the proportions of an ordinary human being, perhaps someone in a fireman's suit, then it remains possible that this was a hoax played on the officer (for a full report, see *The UFO Conspiracy* by Jenny Randles).

Recent history has added one more photograph to our list of aliens preserved on celluloid. And this is a real puzzle.

On 1 December 1987, former police officer Philip Spencer set out in the early morning to walk across Ilkley Moor, in Yorkshire, England. He carried with him a compass and a camera, intending to photograph Ilkley

If this photograph is genuine, does it prove the reality of alien visitors? Former police officer Philip Spencer claims he was startled by this creature while walking across Ilkley Moor in North Yorkshire early on 1 December 1987. He had time for one shot, before it scuttled away behind a bluff. Seconds later, a silver disc shot up into the sky. Spencer suffered a time-lapse, and a skilled clinical psychologist uncovered an abduction account during hypnotic regression. The authors found Spencer to be a very persuasive witness. (*Peter A. Hough*)

town from the top of the moor. At a place called White Wells, rife with folklore associations, something caught his attention and he spun round to see a small 'creature', some 10 feet (3 metres) away. Spencer shouted, raised his camera, and the creature scuttled towards a bluff, then stopped and made a dismissive motion with its right arm. It was at this moment that Spencer pressed the shutter button on his camera.

Then he chased after the creature, which had disappeared around the bluff. Turning the corner, Spencer was confronted with a hovering silver disc which shot up into the clouds and disappeared. Instead of continuing his journey, he returned to Ilkley, where he was confused even further to discover that almost two hours had disappeared out of his life.

The photograph, taken on 400 ASA film with a Prinz Mastermatic 35 mm camera, was underexposed and slightly fuzzy due to camera shake. But it shows a small green figure with ape-like arms and strangely jointed legs. Spencer added to the description by telling us it had large ears and three digits on each hand.

At face value, one might presume this is just another hoax. For instance, the fuzzy picture is what one might expect if a hoaxer wanted to hide what in reality might be a crude dummy of some sort. Conversely, given that the witness had no time to set up his camera for the correct lighting conditions, one might equally have been suspicious if the print had been crystal clear! But this is a very complex case in which we have both been closely involved.

During the intense investigation which followed the alleged incident, we have been assisted by scientists from the University of Manchester, three photographic specialists, including one from Kodak, and a clinical psychologist.

It has to be said that all of our investigations so far have led to ambiguous results. There is nothing clear-cut to prove that the photograph is of a genuine 'alien' creature, nor is there any evidence to prove it is a hoax. Measurements show that the figure was around 4 foot 6 inches (1.4 metres) tall: a sizeable dummy to drag up a very steep incline on to the top of the moors, photograph, then drag back again. Also, Philip Spencer, actually a psuedonym, appears to lack the normal motivations of a hoaxer. This former policeman had made it quite clear that his real identity must never be revealed. At the time he was thinking of reapplying to join the police force, and now holds a senior management position with an American company operating in Britain. He is acutely aware that the release of his identity – the picture has attracted a certain amount of media attention – could ruin him professionally and socially. Most hoaxers want their picture and names in the papers.

Secondly, hoaxers usually show interest in making money out of their claims. Philip has always refused to accept any recompense in connection with the case, even though, at the time, it would have been welcome to him. The incident cuts much deeper than a 'mere' sighting of a UFO. The only hoax scenario which makes sense to us, would be if Spencer were acting as a 'front man' for a group out to test and discredit ufologists. When we put this to him, Spencer just shrugged and said: 'I don't see the point in that. I've got better things to do with my time.'

It came as no surprise to us when hypnotic regression revealed an abduction scenario by 'aliens' aboard the alleged spacecraft to fill in the missing time. We also discovered that the photograph was now 'remembered' as being taken *after* the abduction, which had been wiped from his conscious mind. This would clarify some of the discrepancies noted with the story.

Jim Singleton, the clinical psychologist who carried out the hypnosis, found Spencer to be a balanced, sensitive individual. He is as certain as he can be that Spencer was genuinely under hypnosis. More than this, Singleton stated that Spencer recounted his experience in exactly the same way as his clinical patients who have suffered genuine traumatic experiences.

However, the case stands or falls by the unique photograph. The *Daily Star* newspaper first presented it as a UFO puzzle (without the consent or knowledge of the witness or ourselves). When a complaint was lodged, it subsequently explained the figure away – quite absurdly – as an insurance salesman carrying a brief-case, supplying a picture to prove this. This man later admitted to us that the comparative photograph had been set up by the *Star*. He did not claim that he was the alien! The *British Journal of Photography*, in its February 1990 issue, also poked fun at the matter. All we can say is that the case remains unresolved and that these rather feeble attempts to dismiss it have not helped clarify the truth in any way.

If it is a hoax, it is a clever one. If it is genuine, then it may be unprecedented. We can merely offer the facts, reserve judgement and suggest that you decide for yourselves.

Peter Hough was recently discussing the debate surrounding the picture with its photographer. Spencer, staring at the image on the print, shook his head, and said: 'I just wished I'd taken a clearer photograph. I had no idea there would be so much interest in it.'

20
Alien Properties

If aliens really are visiting our planet, as some people believe, then in the best traditions of pulp science fiction it makes sense that 'they' should be taking samples of soil, carrying out vivisection on our beasts, and abducting members of the human race for mental and physical testing. And that is exactly what the wealth of UFO reports indicates.

In real-life close encounters of the third kind – that is the sighting of a craft and entities – there are many examples of witnesses observing aliens apparently taking samples of soil and grass, or testing the ground in some way. During a series of detailed UFO encounters in the New York area, off-licence manager George O'Barski had an unexpected experience on his way home in the early hours of a January morning in 1975.

It was his custom to stop at an all-night diner and to get there he took a short cut through North Hudson Park. When he entered the park on this occasion, however, his radio started emitting static before cutting out altogether. Suddenly he heard a noise and saw an object, shaped like a 'pancake', drift past him and land in the park. In amazement he watched as about 10 small figures, wearing uniforms, came down out of the object on to the ground.

They held implements similar to small shovels, and, 'working like beavers', started digging up samples of turf and placing them in little bags. The entities climbed back into the craft and it took off. Ten months later, investigators were still able to see 12 to 15 triangular holes in the ground. They had silted up with the rain, but the grass had not re-rooted. Willian Pawlowski, doorman of a block of flats which looks out over the park, described seeing an identical craft, on what was probably the same night.

There have been literally *thousands* of unexplained animal mutilations, mainly in America, Canada, Puerto Rico, Brazil, the Canary Islands and Spain. Most of these have been cattle, although goats, chickens and dogs have been reported too. These animals have been discovered with internal organs surgically removed, and many of them drained of blood – although no blood was found at the locations, as if the work had been carried out elsewhere. In some instances, sightings of unexplained lights have been reported in conjunction with the mutilations. They have been blamed on natural predators and satanists, but the

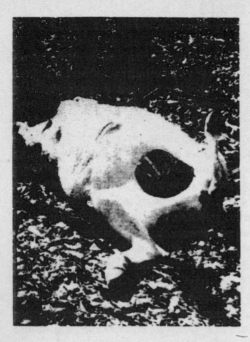

A mutilated cow from Van Zandt County, Texas. Thousands of animals, mostly cattle, have been found across North America with internal organs, tongue, eyes and lips removed with surgical precision. In some cases, blood has been totally drained away, although none was found on the ground. Sceptics and government officials blame natural predators and satanist groups, but many ranchers and local police officers disagree. UFOs have been linked to the killings after sightings at some of the locations. (*Grand Saline Police Dept.*)

sheer number of killings defies such obviously rational explanations. Local police officers and ranchers say that it is impossible for foxes and cultists to have carried out the removal of eyes, tongues, hearts, brains and reproductive organs with such precision.

It is a real mystery, and inevitably one that some believe has a causal link with the UFO phenomenon. Although no human predator has ever been caught carrying out a mutilation, one women believes she saw a calf being abducted by UFO entities.

Judy Doraty, her daughter Cindy, along with her mother and sister-in-law, were driving back from Houston, Texas, in May 1973. It was a clear moonlit night and the family became intrigued by a bright light hovering in the sky. At some point Judy stopped the car, then, after watching the light for a short while, continued home. Afterwards, Judy began suffering terrible headaches and uncontrollable feelings of anxiety. Eventually, she was hypnotized by Professor Leo Sprinkle and she returned to the UFO sighting to re-live the episode. If her hypnotic testimony is to be taken at face value, then it would seem that some sort of editing process had been carried out on the memories of those in the car.

Under hypnosis, it seems that the light came much nearer than anyone recalls, and from then on, the episode becomes even more bizarre, when a calf in a nearby field becomes involved:

'It's like a spotlight shining down on the back of my car. And it's like it has substance to it. I can see an animal being taken up in this. I can see it squirming and trying to get free. And it's like it's being sucked up.'

Judy then feels she is in two places at once – standing zombie-like beside the car and also in a strange room with the calf and 'two little men'.

'It's taken into some sort of chamber. And I get nauseated at watching how they excise parts. It's done very quickly, but the calf doesn't die immediately. There's tissue . . . it's laid out flat and smooth. It glistens. And there's needles in it, or what appear to be needles. It may be probes, I don't know. But it has a tube connected to it. And the same thing with what appears to be testicles. The same with the eye and the tongue.'

The real horror arrives when she sees her daughter being lifted on to an 'operating table'.

'I'm just afraid they're going to do to her what they did to the animal. They put her to sleep I guess. They're just examining her, but I'm so afraid they're going to cut her or something. They don't listen, they just ignore me . . . go about their work as if it's nothing. They don't seem to have any emotions. They don't seem to care. They just take some samples from her . . .'

According to Professor Sprinkle, Judy Doraty is 'a sincere, honest and credible person.'

There are now many hundreds of alleged UFO abduction cases on record. In ufology they are known as close encounters of the fourth kind. About 50 per cent of them come to light through the use of regression hypnosis, although some abductees have pieced together their recollections via periods of missing time, either spontaneously or bit-by-bit over weeks, months and years.

Hypnosis may be a short-cut to suppressed memories, too terrifying for the conscious mind to handle in the normal way. But hypnosis itself is no truth drug and can reveal one's inner fantasies just as readily. The biggest problem for the investigator and witness is to distinguish between the two and for this reason, while many ufologists embrace it warmly, some UFO groups now reject the use of hypnotic regression, claiming it confuses matters and can potentially bring psychological harm to a witness if not expertly supervised.

The case which brought the whole abduction element of the UFO experience to the notice of public and ufologists alike, was that of Betty and Barney Hill from New England.

On the night of 19 September 1961, the Hills drove back from Niagara Falls, over the Canadian border and into New Hampshire. Sometime before midnight they noticed a peculiar bright star-like object, moving in the southwesterly sky. Barney stopped the car so they could observe it

In 1961, Betty and Barney Hill had a UFO encounter followed by a two-hour period for which they could not account. This was the first abduction case which used hypnosis as a tool for memory retrieval. It received much prominence as a result of the investigative and writing talents of John Fuller. One of the best-documented cases ever, the controversy still rages about whether the abduction part of the experiences was a shared fantasy or an actual kidnap by alien beings. (*Jeeves Studios*)

through binoculars. He told his wife that it was probably just an aircraft on its way to Montreal.

They were travelling in a largely uninhabited mountainous region, with no other vehicles on the road. Suddenly the object changed direction and headed their way. Betty focused the binoculars and saw a disc-shaped object with a band of light half-way around its circumference. Still convinced it had to be an aircraft or military helicopter, Barney got out for a better view. His wife heard him repeating over and over again: 'I don't believe it! This is ridiculous!'

The UFO now hovered close by, and he was able to describe several figures watching him through lighted windows. One figure in particular stared down at him and he felt afraid. Suddenly he panicked and became hysterical. He jumped into the car and stepped on the accelerator. Shortly afterwards the couple heard several curious 'bleeps', like the timer on a microwave, from the rear of the car.

It was only later the Hills realized they had no conscious recollection of 17 miles (27 kilometres) of their journey, and two hours of time was unaccounted for.

Betty and Barney began to suffer extreme emotional stress, and Betty in particular had terrible nightmares and strange dreams. After seeing two different doctors over a period of a year, they were passed on to Dr

Benjamin Simon, a prominent Boston psychiatrist specializing in hypnotic therapy. Over the next six months the Hills were independently hypnotized to probe their anxieties, which seemed to be connected with their UFO sighting. However, so much time had passed between event and hypnosis that it was never clear to what extent one person's story was intertwined with another, as they inevitably discussed what took place with many people in the aftermath of the encounter.

For what it was worth, a terrifying story emerged of their car being stopped by a group of strange looking 'men' with pear-shaped heads and large wrap-around eyes. These beings forcibly carried them into the UFO, where they underwent examination. Much of Barney's hypnotic testimony is emotional to the point of hysteria. At one stage, this large man had to be physically restrained by several people.

The case was investigated and very effectively written up by John Fuller in his book *The Interrupted Journey*. The account is highly complex and has been a source of debate for nearly 20 years. Dr Simon came to the conclusion that the Hills were neither lying, hallucinating nor suffering from any detectable psychosis. He told Fuller during their first meeting: 'You are going to have your hands full with this story. There are many things that are unexplainable in this case. I threw many kinds of tests at them during the months of therapy. I couldn't shake their stories, and they were definitely not malingering.'

Yet although Dr Simon accepted the consciously remembered UFO sighting, he was less sure about the hypnotically recalled abduction. In fact there was a radar trace of an unknown object made at nearby Pease Air Force Base suggesting something physical was at the heart of the story. At one point Dr Simon speculated whether this UFO had acted as a catalyst for an even more exotic dream-like experience. He also wondered if Betty Hill's terrifying nightmares might in some way have been transferred to Barney and become closely identified with a real experience – comments taken up by others since.

During the abduction, Betty Hill claimed a needle had been inserted through her navel – some sort of pregnancy test, the aliens told her. Dr Simon told Fuller: 'There's no such medical procedure. This is the sort of thing that makes me doubt the story of the abduction.' But a test was later developed to study amniotic fluid cells in pregnant women for chromosomal aberrations – Down's Syndrome – using the procedure of passing a needle through the navel.

In one of Barney's hypnosis sessions, he describes how a device was put on his genitals to extract semen. This harvesting of genetic material has emerged strongly in modern American abductions, although not, it seems, in the majority of cases from other parts of the world.

The cover illustration to a story by John Campbell, writing as Don A. Stuart. Although the invaders are unlike any description obtained of UFO entities, the scene vividly illustrates modern abduction accounts, almost 30 years before the Betty and Barney Hill case!

Indeed this is one of the more significant discoveries about the phenomenon. While they all share a common base of key elements, they are also tempered by both individual mythically and culturally derived factors. British aliens, for example, tend to be far more refined, and rather human-like. North-American aliens are predominantly small, pear-headed, grey-skinned creatures with a cool, clinical style.

A detailed study of the abduction phenomenon around the world can be found in Jenny's book *Abduction*.

Probably the most familiar British case of an alleged alien abduction, because of its wide coverage in the media, is that of former West Yorkshire police officer Alan Godfrey.

Godfrey was on the night shift when he answered a call at 1 am on Friday 28 November 1980. A complaint had been received that a number of cows were wandering around a housing estate in Todmorden. He drove around the estate but drew a blank. Godfrey's shift was due to end at 6 am, so around 5.10 am he decided to return to the estate in case the rogue herd had materialized.

Leaving the town centre along Burnley Road, nearing the junction with Ferney Lee Road, a brightly lit object on the road ahead caught his attention. Instinctively he thought it must be a bus – perhaps a works' bus with all its light ablaze. But what was it doing, stationary, straddling the road? Instead of turning into the estate, he continued along the road to investigate.

Now he was within 30 yards (27 metres) of the thing, and whatever it was, it certainly was not a bus! Alan remained calm and stopped the car. There was an object hovering just above the road. It was similar in shape to a child's spinning-top, dome-shaped with a row of five 'windows' which contrasted darkly with the brightness of the rest of the object. The lower part was spinning.

He was convinced this was no illusion, but a solid machine of some sort. Indeed, his headlights reflected off its metallic-like surface. Picking up his police accident sketch-pad, Alan made a rough drawing, and later estimated its size at about 20 feet (6 metres) wide by 14 feet (4 metres) high.

A 'feeling' overcame him that he 'shouldn't be there', that it 'wasn't for my eyes.' Not surprisingly, he felt safer in his car. But as he tried to radio the police station, a burst of static issued from the unit. This in itself was not unusual, as the hilly terrain often interferes with radio reception.

Then it was gone. One moment it was there, then it had vanished as completely as a burst soap bubble. Yet, the car was now 100 yards (90 metres) further up the road and he had no knowledge of how it had arrived there.

Godfrey drove back to the station, where sceptical colleagues agreed to return with him to the location. Earlier in the night it had been raining but the area of road where he claimed the UFO had hovered was dry, and there was a swirled pattern on it. During questioning, UFO investigators

Police officer Alan Godfrey's boot was found to be cracked right across the sole after his alleged UFO encounter in 1980. Does anecdotal evidence like this go any way towards proving an objective reality of the phenomenon? (*Jenny Randles*)

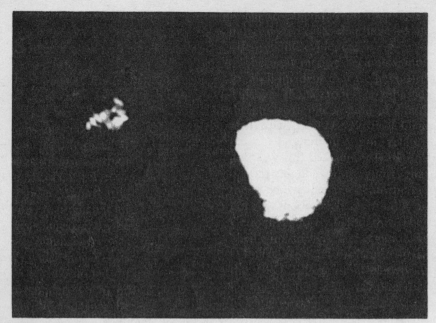

Many ufologists now propose a radical theory for even the closest of encounters. Unusual atmospheric phenomena – which we can prove *do* exist – might be capable of interaction with the consciousness of susceptible witnesses and trigger visions of alien visitors. This atmospheric ball of energy was filmed at Hessdalen, Norway, apparently one of the most UFO-haunted spots in the world! (*Project Hassdalen*)

concluded that Godfrey could not account for approximately 15 minutes of time.

Under hypnosis, his account was occasionally as dramatic as that of Barney Hill's. He found himself inside a 'room' with a tall humanoid male figure wearing something similar to a white sheet. Godfrey was then aware of eight small beings, which he described as 'horrible'. They stood 3 feet (about 1 metre) tall, with heads like lamps! The taller figure 'told' him not to be afraid and he felt the others were 'robots'. Eventually Godfrey was led across to a bed in the middle of the room, which he climbed on to without wishing to do so. What followed was a frightening physical examination. Afterwards he suddenly felt sick and was aware of a pungent smell. Flashing lights filled his head, inducing waves of nausea. Then he was back in the car – the UFO had gone.

Godfrey has always been adamant that what he saw on the road was beyond doubt a physical reality. Everything about the case suggests that this was true. Something reflected the headlights of his car, but that need not have been a spaceship. Other possibilities must be explored as well, such as novel atmospheric forces creating phenomena that science has yet to understand.

As for the abduction experience, despite the fact that after the events the sole of one of his shoes was split right across the middle – Barney Hill's shoes had been scuffed, tying in with his description of being dragged by the aliens – Godfrey is commendably honest, pointing out that he read UFO stories between the sighting and the hypnosis sessions months later. He acknowledges this could have coloured what he said in an altered state, which might therefore be open to other interpretations. While nobody can prove what happened one way or another, if the witness himself is unsure of the objective reality of the abduction phase of his story, we must be wary of forming earth-shattering opinions about extraterrestrial life.

Are aliens visiting the earth, taking soil samples, disecting animals and abducting human beings? This *is* a rather clichéd science fiction scenario, which might in itself be significant. Life imitating Art, or Art imitating Life? Undoubtedly these experiences tell us that something extraordinary is going on, but extraterrestrial visitors are only one possibility. Perhaps the answer is even more complex.

Alien abductions are undoubtedly a major puzzle for science to resolve. However, it is unclear where the answers will come from. Indeed, extraterrestrials may be a welcome alternative to a far more shocking truth: that the spaceships emerge from our minds and that the 'aliens' haunt the corridors of inner space in our heads.

21
Back in the USSR

Evidence that aliens have visited the earth is a regular feature in the Western media. We have become hardened to such tales, learning to regard them as amusing at best and gullible at worst. How many of us would change our views about the universe after reading the latest story in sensationalist newspapers like the *Sunday Sport* in Britain, or the *National Enquirer* in America?

However, on 9 October 1989, something quite remarkable happened to seemingly change all that. Suddenly, the source of information about an alien contact was no alarmist British or American tabloid, but the conservative Soviet news agency Tass!

The story that Tass announced, as mundanely as if releasing the latest monthly returns of the potato harvest, was, to say the least, staggering. A spaceship had – it told the world – landed in the USSR and weird alien creatures had stepped out of it. Absolute proof of this had been established by scientific investigation, Tass claimed.

Here was the moment for which we had all waited. This book has debated in depth what often looks to be solid, but is ultimately rather flimsy, evidence for alien visitors. Now the questions and the speculation had been made redundant by this news from Eastern Europe.

Or had they?

According to Tass's initial reports, this is what supposedly took place. The location was the industrial city of Voronezh, which is situated some 300 miles (480 kilometres) south-east of Moscow. The date of the event was 27 September; the time about 6.30 pm. Dusk was descending after a warm and pleasant day. The first reports left many vital pieces of information out.

It seems that a whole crowd of people in a local park had seen a 'ball of red fire' circle the area and then land. Looking a bit like a banana, the craft had disgorged a few aliens – one, two and three were the numbers quoted. These were over 10 feet (3 metres) tall, with tiny heads. At the landing site, some traces and strange rocks were left, which a local scientist from a geophysical station said were not indigenous to the earth.

Quite predictably, this revelation had major repercussions outside the USSR. Indeed, we now know that it was treated far more seriously in Britain and America than it ever was back in the USSR. Presumably the reason for this was its apparent uniqueness and the fact that while tales of alien landings are amusing and unimportant when they happen on your doorstep, they are potentially dramatic when they escape through the normally rigorous media outlets of the USSR.

Indeed, one Tass correspondent, when challenged on British TV about the veracity of the accounts, seemed almost personally affronted and assured the interviewer: 'Tass never jokes!'

Proof of everyone's belief in Tass's reliability was shown by the astonishing media interest in the story. In America, newspapers like the *Washington Post* and the *New York Times* reported the event in a way rarely accorded cases in their own land. By the fifth day of the furore, the *Times* even carried an editorial stating that this could be 'the story of the century', adding that one could all too easily be too sceptical of such matters.

In Britain the pattern was repeated, with bombardment of stories from all the responsible newspapers that either shun or put down far more credible UFO cases at home. In addition, for the first time since the UFO phenomenon began to be reported, all four British TV channels carried items on their news programmes on one day. In these ufologists from the British UFO Research Association, including their journal's editor, Mike Wootten, and ourselves, appeared to discuss the matter. They had been invited as serious commentators and were found responding to rational questions and debating the political consequences on programmes such as the ITN's *News at One* and the BBC's *Newsnight*.

Why did this case merit such an avalanche of credibility? Almost certainly the rush to judgement was premature, as is shown by the rapid way in which it vanished without trace. The truth – or rather, as close as we are ever likely to get to the truth – was rarely, if ever, featured later and most people were left as confused as ever about what really occurred in that Soviet park.

We have been able to piece this story together, via original press releases, correspondents in the USSR, original articles and other material. The picture which emerges proves many of the points we make in this section of the book.

Firstly, it is important not to see this story in isolation. While to media sources in the West it may have appeared as a sudden and dramatic revelation, it was no more than part of a process that began even before Gorbachev had liberalized thinking and press coverage via glasnost.

The Soviets have long had an interest in UFOs and had announced the formation of a scientific investigation team in 1984, before Gorbachev took power. This followed some disturbing incidents where UFOs overflew secret airspace and shocked pilots. There are even grounds for believing that the terrible mistake in shooting down a packed Korean Airlines jumbo jet, when it strayed into restricted airspace over Sakhalin Island in 1983, may have been UFO-inspired. Only a few months before this disaster, a similar event occurred over Gorky and the UFO had been allowed to fly on unhindered, although it was almost intercepted because it failed to respond to radio warnings. One can see how an identical situation over Sakhalin might have later provoked a tragically more belligerent reaction.

After these events, the Soviets began taking UFOs much more seriously than the Americans were officially claiming to do. Cases flooded in to the authorities on a regular basis, some of which were reported by Tass, but they were rarely picked up by the Western media. The incidents that filtered out to the West via various sources included an even more extraordinary event than the one at Voronezh. This had happened in 1986 when a UFO was allegedly seen to crash into a mountain called Hill 611, near the village of Dalnegorsk, in the Pacific region of the USSR near Vladivostok.

The Dalnegorsk case is like a modern-day Tunguska. The UFO was said to resemble a 'flying sphere' and it hit the mountain and exploded, all but disintegrating in the blast. Nevertheless, the damage was evident and researchers from the Academy of Sciences reputedly found some strange meshed material and bits of glass possibly fused by the heat of the impact. An unknown form of chromium alloy was traced and the debate still rages among Soviet scientists as to what this object was. Some favour a natural but previously unknown type of plasma and point to repeated sightings of UFOs in this hilly area, known as Primorskiy Kray. Others insist the device was controlled – i.e. a small space probe.

The Voronezh landing must be seen in the context of the Dalnegorsk case and others involving aliens. At least one of the reports in the summer 1989 outbreak featured the same sort of alien beings later claimed to have landed at Voronezh – a type otherwise virtully unheard of even in UFO records. So it is not true to claim that the USSR was uncontaminated by Western media stories about aliens or UFOs. It is a nation with a strong tradition of science fiction in both literary and cinema form. Although, as we shall see later, general stories are one thing, but detailed accounts which correlate with one another are a different matter.

One of the points that we made during our media interviews is that glasnost works two ways. While it is true that it allows important news to be commented upon freely when previously it may have been censored, it also creates competition between the Press, which is subjected to some of the pressures of its Western counterparts. As a consequence of the increasing need to feature interesting stories, many are printed without

The site on Hill 611 in the USSR where eyewitnesses described the impact of an unknown object in 1986. Scientists think the object was a meteorite or a comet. Some commentators have other theories.

being checked out thoroughly. A newspaper can rarely wait to authenticate a newsworthy incident. In doing so, a rival may carry it first.

Again, it has to be remembered that from our own experience in these matters, stories that make major headlines and flash around the world are often the result of ufologists going to the media with the 'exclusive' news in the first place.

We saw this happen in Britain with a case similar to the Voronezh one. A UFO was reported by both civilians and military personnel outside a USAF/RAF base in Rendlesham Forest, Suffolk. It reputedly landed, left traces in the forest, radiation levels in a triangular set of indentations at the site and created a mild degree of panic in official quarters. That was in December 1980, but it never made it into even the local media, despite publication in UFO magazines and articles in two news-stand journals. Then, in October 1983, ufologists chose to take it to the British tabloid newspaper, the *News of the World*, from where it spread like a disease all over the world and eventually saw the light of day in many countries in garbled form.

In the Voronezh case, the original Tass correspondent investigating the incident, was (according to magazine *UFO Universe*): 'Vladimir Lebedev . . . who is also a well-known Soviet ufologist [who] . . . has maintained

an active correspondence with Western researchers [such as] ...
Timothy Good in England.'

From this information we begin to see that the story reached Tass in a
very similar manner to most UFO blockbusters, for example, the 'Shuttle
Crew met aliens' saga that had filtered out some six months earlier.

While we have no reason to believe that this was anything other than
an example of a sincere ufologist with an extraordinary tale to tell, it is
worth bearing in mind that at times the Soviet authorities may have had
other motivations for allowing UFO yarns to develop.

For example, there have been major stories about space invasions of
Moscow and 'giant jellyfish' in the skies. These have been successfully
shown by space researcher James Oberg to have been military rocket and
satellite launches. The authorities allowed (and even encouraged) UFO
researchers to promote these sightings as something strange to serve as
disinformation: nobody would take the sightings seriously if they were
disguised as UFO landings.

There are some people – ourselves included – who suspect that this is
not an idea limited to the USSR. Cases such as the 'UFO crash' in
Rendlesham Forest (or, as we noted earlier, the one at Roswell, New
Mexico) could be examples of military secrets hidden under the smoke-
screen of a UFO incident. Once developed, it allowed the authorities to
kill off all serious investigation of the matter. Nevertheless, at Voronezh,
the basic facts appear to be that something anomalous was experienced.

The second set of Tass reports on the story (carried less often by the
Western press 24 hours later) gave more details. By adding data from our
own sources, we can now report what supposedly happened.

It seems that there were sightings of lights and other things in the skies
about the city intermittently during September and October, and one
report of a sky watch at the site involving over 350 people waiting,
unsuccessfully, for the UFO to return. Obviously, social factors could
account for some of these sightings – ordinary objects, such as aircraft
and weather balloons, could have been wrongly identified. We have the
reports of these sightings and they contain entries like the following: on
23 September, a 'round ball of red with a black door' out of which
emerged a 'robot', according to 12-year-old Roman Torchine; on 26
September 'a figure' with a small head and 'three spots of different
colours on his chest', according to another child from the same class as
Roman. Dr Henry Silanov, Director of the Spectral Department of the
Voronezh Geological-Geophysical Laboratory, and investigator of the
phenomena, contacted us with this information:

In the period between September 21 to October 28, 1989, in the Western Park in
Voronezh, six landings and one sighting (hovering) were registered, with the

appearances of walking beings. We have collected a wealth of video materials with eye witness accounts, particularly from pupils of the nearest school. We have no doubts that they are telling the truth in their accounts, because details of the landings and other signs are recounted by the children, who could otherwise have only got this information from specialist UFO literature which is not publicised in our country.

We have also been able to acquire copies of statements made by schoolchildren involved in several of these close encounters of the third kind. This is what Vasya Surin had to say:

It was a few days after I had come out of hospital on September 19. We were playing football in the park during the evening as it was just getting dark. Then we saw from the direction of the [untranslatable name of a building] first a sort of pink haze, as if someone had lit a bonfire in fog. This haze was higher in the sky than the chimney stack of the building. Then we told everyone who was around, and they stopped to watch. From the middle of the fog appeared a sort of red sphere. The fog did not move, it stood still, but the sphere flew away at great speed.

It stopped over a tree, then lowered slightly on to the top of the trees, rested on it and the tree bowed under it. We got frightened, we thought they might see us if we ran home and might take us away. Then we hid in the bushes and kept watch.

A door opened in the sphere while it hovered over the poplar tree. A person (being) looked out. He was tall – about three metres [10 feet], shone silvery, and his arms were down to his knees. He stood up like this, looked around, but could not move his head, it was as if he didn't have one, as if it was just a continuation of his shoulders – with three eyes. Two at the sides, and a third just a little higher up, which moved up and down, left and right. 'He' looked down and the door

These four children witnessed the sightings at Voronezh during September 1989: (from top left to bottom right) Lena Sarokina, Vasya Surin, Vova Startsaev and Aliosha Nikonov.

closed, and the sphere passed down a little more on the tree. Then the door opened and they started to climb down the tree. First a robot then an extraterrestrial. The robot had no head. At first we thought he was an extraterrestrial too, an extraterrestrial walked up to the robot and seemed to punch out some numbers to enter a code on the robot's chest, and it started to walk mechanically.

It walked along the road a bit, while the extraterrestrial said something and a rectangle appeared in the ground. The rectangle was red and you could see lights in it. Then he spoke again and all this disappeared. One of the little boys who hadn't managed to hide, shouted out in fright. The extraterrestrial looked at him and the boy seemed rooted to the spot. When the extraterrestrial looked at him, his side-eyes seemed to light up. Only when he stopped looking at him, did the boy move a little.

A young man was walking along the road, perhaps to the bus stop. The extraterrestrial noticed him, climbed up the tree into the sphere and came out with something like a pistol, yellow with a telescopic sight. The young man, when he saw the alien, ran off and the latter aimed the pistol at him and he disappeared. The robot walked up to the extraterrestrial who said something to it, before climbing the tree back into the sphere, whereupon it flew away. As soon as it was gone, the man re-appeared and carried on walking as if he had seen nothing!

After a few minutes some kind of 'saucer' flew towards us which stopped next to the poplar tree. It had little 'circles' in it which issued fire and a hissing sound. Two aliens came out of the saucer and went off in the direction of some electricity pylons.

They were the size of an ordinary man, their faces were wrinkled. When one of them went up to an electricity pylon you could see he looked as if he was burning. He burnt completely. The saucer then flew away.

Several young witnesses signed statements to another alleged incident, which occurred at approximately 2.30 pm on 28 October. Vova Startsev was playing truant from school with his friends when they saw a large pink sphere overhead with the symbol Ж painted on it.

It was flying quite low, it passed the street lights and landed here. It was pink but kept changing shade. On the left-hand side of the body were two antennae. It pushed out four legs, a hatch opened, a ladder came down and two beings and a robot came out. They carried the robot, set him on his feet, gave him artificial respiration then he walked like a man. It came up to me, followed by one of the extraterrestrials. He was just under two metres [6 feet] tall, stretched out his hand towards me, but I ran towards a tree and climbed it, shaking with fear. The alien had a big head, twice as big as ours, and three eyes in a row.

Vova was joined by one of his friends, Sergei Makarov. He said the aliens wore silver suits, silver waistcoats with silver buttons and boots. Their faces were the colour of 'grilled beefburgers', but the skin was smooth. When the door had opened in the object, a blinding light from inside stopped them seeing any interior details. As the pupils

The strange symbol like the one reported to be on the craft at Voronezh. It has also appeared on documents, reported to come from the planet Ummo, sent to ufologists around Europe. This particular photograph, allegedly of an Ummo spacecraft, was taken near Madrid. Computer enhancement showed what some analysts suspected might look similar to a piece of fine thread or wire suspending the object.

watched, the legs retracted into the object, then it hovered before rising and flying off.

It seems too that this same object, with the same curious sign on it, was observed earlier, towards the end of September, at just after 7pm, by a young man called Denis Valyerervich Murzenko.

He had changed clothes to attend a concert with his mother, and decided to go for a walk while she was getting ready. Looking up into the sky he saw something pink, shaped like an egg, with rays of light coming off it. The object came closer and began to swing from side to side like a falling leaf. At this point two supports came out from underneath, and Denis saw the same curious sign painted on it, and the outline of someone in the object. He was very frightened but this is what he described:

The 'person' seemed to be about four and a half feet [1.4 metres] high, with an old face. I stood still and it kept coming down lower and lower. I became frightened and ran off. When I turned I saw bright beams of light. The object stopped next to the flagstones. There was some kind of sound like music. In the hatch I saw a silhouette of a person.

But an added bizarre touch was to occur several days later. Denis noticed a suspicious man hanging around at the location of the UFO landing. Denis went up to him and noticed a metal plate on his chest with some marking on it.

He stood for a while, got on his knees, touched the grass, then started to wave his arms about. Then he got up, ran off, then walked normally I followed him and he went into a shed. I was only twenty metres [22 yards] behind him, but when I arrived he had gone.

The Voronezh landing also prompted people to report things that would have otherwise gone unreported, thus creating the idea of a wave of sightings. Most of them could be put down to plain misperception, but in the reports mentioned above, we have something more than mere sightings of anomalous lights: here we have detailed descriptions of objects and interractions with aliens.

The reports are fascinating. Of course the ones we have available for study have been made by schoolchildren, and hence would be dismissed by many people. But as our attitudes have been forced to change over the reliability of children when reporting abuse, perhaps we should be more open minded here, too.

Having said that, the reports are fantastical, and remind us of the type which issued from America in the late 1960s. A few things are especially noteworthy, considering, as Henry Silanov says, that the Soviet people are starved of 'specialist UFO literature.'

The curious symbol noted in at least two of the cases is virtually identical to a sign attributed to the Ummo. Ummo is supposedly a planet where aliens, domiciled on earth, come from. These aliens have been contacting UFO researchers and academics, mainly in Spain, since 1965. Photographs have been produced of a spacecraft bearing the symbol, which has since been proved a fake. The figure is also an astronomical glyph for the planet Uranus, and, perhaps more important in this case, very similar to a letter in the Russian alphabet!

The pink mist which the Voronezh object appeared from has been described in many cases in various parts of the world. The 'falling leaf motion' of the object, too, has been noted by many people. Witnesses also have informed investigators of strange 'men' who have appeared at UFO sighting locations days or weeks later.

In our search for the aliens, we are not suggesting that you take these accounts at face value, but nor are we proposing that they have been invented either. It may be the best example yet of the fantasy/reality incongruity of the subject. But, adult witnesses were surprisingly hard to come by, given that they must have been there or in the vicinity – at a nearby bus stop and in the high-rise flats overlooking the park. A visit by Soviet TV on 11 October failed to find a single adult who would describe seeing the landings. This was also the experience of one British journalist we interviewed who had visited the area.

One adult who was a witness is Sergei Matveyev, a lieutenant in the local militia who seems to have been known to the children who made the main reports. He saw 'an object almost 50 feet [15 metres] in diameter over the park' that same night. He thought it was an illusion, because he was tired, but it remained there after he rubbed his eyes. Nevertheless, he did not see it land and did not see any aliens or 'robots' either.

However, Viktor Atlas, mayor of the city, announced his faith in the children who did claim these things. They were aged between 13 and 16 years old and had mostly been playing football at the time of the main incident.

But why did Tass regard the Voronezh incident as significant enough to publicize? The alleged physical evidence of an alien landing appears to be the reason.

Tass quoted Dr Henry Silanov: 'We found two mysterious pieces of rock. At first glance they looked like sandstone of a deep red colour. However, mineralogical analysis has shown that the substance cannot be found on Earth.' He added, 'Although additional tests are needed to reach a more definite conclusion.' But some reports missed that part out.

Within 24 hours Silanov was flatly repudiating the claim. He had been misquoted, he insisted – not uncommon when UFO matters are reported. In fact the rock was nothing but haematite, an iron-rich ore.

Tass also reported that peculiarities at the landing site (which was only vaguely identified by the children) had been 'scientifically investigated' nearly two weeks after the incident by the use of a method described as 'bio-location'. This turns out to be a form of psychic dowsing that reputedly found magnetic 'hot spots'. Few people would describe this as a scientific field investigation and so again the story tended to mislead by implying that tangible evidence had been discovered.

A report on 13 October claimed that a site analysis had 'registered incredibly high levels of magnetism. It is evident that something produced it.' This was Dr Silanov again – a quote not retracted. Imprints supposedly left behind were measured by someone from his laboratory who concluded that the UFO must have exerted a pressure of 11 tonnes.

Finally, on 18 October, Ludmilla Makarova from the local militia investigation unit commented on their work at the site. There was 'abnormally high radioactivity' throughout the park. She added, 'I don't know what caused it, but it is certainly an increase'.

By now the fame of the location had encouraged a member of the new breed of entrepeneurs to jump on the band wagon. A tour was being run to the 'extraterrestrial landing site' for just 59 roubles. If you were willing to add another two on to the bill, you could converse at the spot with 'specialists' who knew all about such things.

However, as the case descended into a morass of legend and confusion within the USSR, and oblivion elsewhere, one final investigation was conducted by the local university. Igor Sarotsev, in the spirit of debunkers everywhere, advised on 28 October that 'There were no extraterrestrials at Voronezh.'

But hold on – what of the radiation? Interestingly, this was at its strongest inside the shallow indentations found within the ground and so it was presumed to have been left by the spaceship. This was exactly what was discovered in Rendlesham Forest after the 1980 'crash landing.' Here the sceptics insisted that the holes had been dug by rabbits, although no such solution was put forward at Voronezh. Holes naturally accumulate debris, leaves from trees and rain water, for instance. If any radiation is in the atmosphere, these will help to concentrate it into the depressions.

Sarotsev was absolutely convinced the readings that everyone had obtained were not of intergalactic proportions. He insisted: 'We have found no anomalies in the ground or vegetation. The presence of a large quantity of caesium [a radioactive isotope] does not constitute sufficient proof. After the Chernobyl catastrophe this sort of thing is found in many places.' However, the Rendlesham Forest incident was six years before the Soviet nuclear reactor caught fire, contaminating much of Europe.

So, the Voronezh affair disintegrates into a mass of ambiguity, despite the incredible reaction of the Western media and the apparent initial strength of this story. It is no more proof of extraterrestrial visitation than any of our other cases have been. Yet it is proof of *something*.

Once again the aliens that we so dearly seem to want to contact are proving both frustrating and elusive.

22
Where Are They?

In his series of novels beginning with *Helliconia Spring*, science fiction author Brian Aldiss speculated about what life would be like on a planet where the orbit is so complex it takes centuries to pass from season to season. That is the fate of worlds around suns with companion stars – parts of double or triple star systems.

It is by no means certain that planets could survive in unstable orbits under such circumstances, or that any kind of intelligent life could develop in the resulting environment.

This is a serious problem, because over half the stars similar to our own sun are part of such binary systems. As stars like our own sun are the only type we feel confident will generate life, their nature poses yet another major restriction on how many neighbours we may have. Indeed, some astronomers are beginning to wonder if the odds are not stacked so heavily against other life forms that we could be utterly alone after all. If that is true, then our search through the evidence in this book and the quest to establish contact by scientific means will never produce anything except perhaps some data about human belief systems.

That might explain why this section of the book has been almost depressingly negative. Our objective analysis of the claims for alien contact have more often than not brought conclusions that fail to provide any real proof, and at best only tantalizing possibilities. We have searched for alien civilizations who may be in contact with us, but we have found very little sign of them in terms that would persuade the hardened scientist.

Jerome Pearson, a space engineer with the US Air Force, referred to 'the lonely life of a double planet' in his article for *New Scientist* (25 August 1988). He pointed out that the near-unique way in which the earth and the moon form a double-planet system may also have contributed to our specialness and that this would not recur very often in the universe.

He notes that, as long ago as 1939, the great physicist Enrico Fermi asked the question: 'Where are the extraterrestrials? Why haven't they landed?' Assuming that we are forced by our evaluation of the UFO evidence to agree that they have not, this is a very valid and disturbing problem.

The military base at Canaveral where signals from space are detected and unmanned rockets are launched. (*Jenny Randles*)

As Pearson says:

If life arises naturally, then we might expect thousands of living planets and perhaps hundreds of civilizations in our galaxy . . . [Yet] there are no signs that they have visited us, are on their way, or even that they are communicating among their many settled worlds.

We have already discussed some of the possible reasons for SETI proving so fruitless, but as each new search continues and finds nothing of substance, we are dragged toward the disturbing conclusion that something may well be very wrong with even the most basic of our assumptions.

When the *MUFON Journal* reprinted Pearson's article in America, it provoked some response, to which the scientist graciously replied. He said in the May 1989 issue:

Personally, I am a great fan of Isaac Asimov's galactic empires and I long for the excitement of extraterrestrial contact . . . But 30 years of SETI have shown zero results, and we must therefore entertain the notion, however unpleasant, that we are either alone in the galaxy or we are among its most advanced species.

There is a simple but profound reason why he makes this statement. It has been phrased in differing ways by many astronomers, but no more poetically than by French science writer Aimé Michel (*Flying Saucer Review*, volume 25, number 6). Michel uses the term 'the cat-flap effect' to make his point: 'The first civilization to cross the cat-flap point explodes, literally, into Galactic space, and occupies it totally within a negligible period of time . . .' Negligible being on a scale of millions of years, which is indeed nothing in comparison with the billions of years of history which the universe already has. In this way, Michel suggests we can forget all those complex equations to work out how many suns have how many planets with how many intelligent life forms growing on them, and ask just one simple question: 'Has anyone else yet crossed the cat-flap?'

If nobody has, then either we are totally alone or are the most advanced species in the universe – the very first intelligent race (just as Pearson speculates). Both of these ideas are unpopular and generally thought so unlikely as to be practically impossible.

Yet, if we are neither the first nor by any means alone, why has nobody crossed the cat-flap to leave their undeniable mark all over the galaxy?

Michael Swords, Associate Professor of Natural Science at Western Michigan University, has had a stab at resolving this dilemma. In his article 'The Third Option' (*International UFO Reporter*, January/ February 1987), he reminds us that there is an alternative to 'they don't exist versus they do but can't get here' argument. This is that they do exist, can come to our system but either have not bothered to do so, or *have* come and are deliberately hiding their presence.

In order to make his point, he rattles off a series of theories which fit in with this third option via 'decision-trees'. Effectively these ideas boil down to the following possibilities:

1. Aliens develop better things to do and never travel or try to establish contact in other ways with other races. This seems implausible as surely some civilizations would have wanderlust intentions and it would need only one who was close enough to cross through the cat-flap.

2. Aliens have travelled through Michel's cat-flap to communicate but we were missed by pure chance or because there is nothing special about our sun. This has some persuasiveness, but it will probably only suffice as an explanation if there are very few civilizations in the universe, because if there are many, the odds that we have been missed would be negligible.

3. Aliens did explore, via the cat-flap, far back in our past, but we were of no real interest as a race and so we were ignored. Again this is feasible, especially if we are dealing with truly alien mentalities who may not share our definition of what is interesting. For example, suppose to them the only interesting civilizations are those who have green hair or can sing songs in the frequency range far beyond the human level, they may have taken one look at us, concluded we were an irrelevance to the universe and gone on searching elsewhere. However, this idea also seems to require a limited number of alien races, as otherwise surely one of

them would have found us, considered us moderately interesting and so retained some kind of monitor probe down through the ages, which, of course, might involve one return visit to see what we are up to every few million years, with the next landing here due on Friday week!

Michael Swords extends his last idea further, suggesting that aliens may have visited and been interested, but decided to leave us well alone. Remember the warning of Zdnek Kopal about how a civilization might choose to destroy us before we infested the universe like a plague of rats? Other possibilities are that a giant quarantine sign has been set up, warning that the earth was full of hostile, primitive inhabitants, and that any self-respecting alien intelligence should steer clear! This would explain why we are getting no replies or having no visitors from space. It is part of a dastardly plot to shun the earth which has, in effect, been decreed to have the galactic equivalent of body odour!

We have added our own thoughts on Swords' basic premises here, merely to illustrate that there are ways in which the seemingly insurmountable difficulty of the cat-flap effect can be overcome. These are only a few of the suggestions for accounting for our lack of evidence.

However, it is fair to say that Professor Swords thinks that these ideas are less likely than the possibility that the cat-flap *has* been breached, the aliens have visited earth, and still do, but there are reasons why their presence here is less than obvious.

Of course, given the fact that Swords is one of a growing band of

Crowds gather at Port Canaveral during one of the earlier Shuttle launches in November 1983. The lure of space travel remains despite more than 30 years of manned space flight, such missions being almost routine today. (*Jenny Randles*)

scientists (especially in America) who are very knowledgeable about the UFO phenomenon, this is not too surprising.

He suggests that perhaps the aliens obscure their presence for one of two basic reasons. Either their motives for being here do not include us, indeed we may even get in the way of their experiments if their existence were established, or else they are running a long-range project to educate mankind which works best without direct intervention. In this way, they dangle carrots on a stick for us to follow, stage mental sideshows and play games with human society in the hope of teaching moral or ethical lessons that will eventually improve our race to the point where open contact might be a practical proposition.

In Swords' opinion, either one of these possibilities best explains the UFO evidence, which, especially in America, is widely considered most likely to be extraterrestrial in origin. He seems to prefer the solution that 'aliens have their own motives', because it better seems to fit the clinical, almost dismissive nature of the abduction data. But it may not explain the apparent non-physical reality of these stories, as has been found outside America.

The American abduction material is not necessarily reflective of the global picture. There is a widespread and understandable provincialism, when considering the UFO phenomenon which is not unlike the anthropomorphism to be found when science offers speculations about the behaviour of aliens.

Whilst Swords' set of ideas about the third option is fascinating and potentially illuminating to our research, it also suffers from these nagging doubts that we may be giving aliens attributes that are really nothing more than human traits in thin disguise. Surely any visiting extraterrestrials would be totally alien, probably to the point that we would have no way of deciphering their motives, let alone psycho-analysing them?

Others have also bravely tried their own methods to account for the apparent lack of alien visitors to our solar system. Raymond Reddy said in *MUFON Journal* (May, 1986): 'Life may not have begun elsewhere so very long before it began on Earth, and even if it did, it may not be all that far ahead today, biologically or technologically.' He is talking about economic, social and other pressures which may slow down progress once a certain level is reached. Another school of thought is that space flight may be achievable by any race with technological skills, whether it has intellectual superiority to ourselves or not. Supporters of this approach envisage a real-life version of the comedy science-fiction film *Morons from Outer Space*, where a group of litter louts and amiable but very dim extraterrestrials from another planet land on earth. They *use* the advanced technology of their civilization but they have no idea how it works. This makes sense: after all, how many of us understand the workings of a TV set or car engine, yet most of us use them on a daily basis?

The purpose of these comments is to show that aliens may be here and behaving in a manner we would not predict. Just because we might plan to fly to Alpha Centauri, discover any life that might be there and establish open and friendly contact does not mean the same thing would happen in reverse. The absence of alien visitors could be a result of nothing more than our inability to realize they are here. Perhaps they are too *alien*, too covert, too stupid, or too wise to mix with the likes of us!

Also, in the history of the cosmos, a few thousand years is nothing, so it may also be true that if there is substance to any of those Ancient Astronaut claims, this answers our question about where the aliens have gone. As we saw, there are a few vague hints within the data that earth may have had contact with aliens in the past, even if most of the evidence is more likely to be a demonstration of the clever skills of our ancestors rather than signs left behind by their extraterrestrial mentors.

One man who believes we ought to take this possibility much more seriously is Dr Vladimir Rubtsov from Kharkov in the USSR. He told us in a letter: 'I have been studying the problem of extraterrestrial intelligence during the last 20 years with particular attention to fringe topics such as UFOs and paleovisits.' This latter is a scientific term he wants us to use for alien visitation in the geological or historic past. In a piece entitled 'The Problem of Paleovisits' published by the *Journal of the British Interplanetary Society* (volume 36, number 11, 1983), Dr Rustsov ponders the apparent lack of evidence for ancient astronauts and says: 'This question has been raised usually together with the overt or ulterior assumption that science has carefully studied all the data bearing on it and has come to the definite – strictly negative – conclusion.' He disputes this and considers that it has not been adequately discussed in a scientific fashion.

Rubtsov is dismissive of much of the work done by what he calls 'many well-meaning but poorly-trained amateurs'. He notes that serious consideration of paleovisitors is widespread in the USSR, but that Western scientists tend to show no interest, despite the fact that they must be aware how the discovery of any such evidence would resolve the paradox of where the missing aliens have gone.

He warns that it is dangerous for science to be seen as opposed to paleovisits whilst non-science is in favour of them, as this must have a negative influence on those who are unfamiliar with the evidence. It may cause them to presume that the data have been assessed as nonsense by science, whereas in effect the facts have actually not been assessed at all.

The NASA Kennedy Space Center at Cape Canaveral, Florida, where the human race is reaching out for the stars and hoping for alien contact. (*Jenny Randles*)

It is interesting to note that Rubtsov is equally outspoken on the question of UFO evidence. He sees a double standard between science pouring money and thought into the theoretical debate about whether aliens could contact us, while rejecting any examination of UFO data 'by slighting the problem as an "unscientific" one', despite no more than a cursory look at the reports. He displays an eminently sensible approach by merely requesting a fairer hearing for the evidence and adds: 'Naturally, it is not excluded that such a serious study will really lead us after all to the conclusion that there are no extraterrestrials in the solar system.' But he says that this is only an assumption at this stage, because science is in the main making up its mind before it looks at the facts.

Many times in this book we have mentioned UFOs. We expect that many readers do not think they merit consideration in the matters we have discussed. In fact, this view, although understandable, results from our conditioning – as Dr Rubtsov suggests – to regard UFOs as a mysterious problem when the search for alien intelligence is a purely scientific one.

However, any reading of the best literature in the UFO field will confirm several things. There is a lot of nonsense spoken and many spurious sightings of no more than everyday objects. However, there is also a good deal of sensible debate and a surprising number of scientists and academics from many walks of life who are devoting their time to the evidence.

The tiny capsule inside which mankind sailed the hostile environment of outer space in early missions. (*Jenny Randles*)

While UFOs are often linked automatically with extraterrestrials, as if it is essential to believe in one in order to accept the other, that is not necessarily the case. A number of leading ufologists reject the ETH (the ExtraTerrestrial Hypothesis), because, after studying the data, in their view it fails to fit the facts. We have seen that this includes people of the calibre of the late Dr J Allen Hynek.

Nevertheless, although it must be understood that many UFOs certainly have nothing whatsoever to do with alien visits – and it is of course possible that none of them do – it is perfectly valid to study the evidence and the speculations, because the UFO phenomenon undoubtedly exists on a major scale and represents data that cannot be ignored if one is tackling the question of contact with aliens on an intellectually honest level.

Let's briefly examine the views of two well-qualified scientists who have spent years researching the evidence and see why they have reached quite different conclusions.

Both accept that something presently unexplained is going on. Indeed it is fair to say that, with almost no exception, scientists who have devoted time and effort to this subject, and made first-hand investigations of the better cases, have reached this same conclusion. The ones who are denying the scientific potential of the phenomenon are, in the main, those who have not really studied it beyond a cursory glance at the usually sensational reports in the media.

Dr Pierre Guérin was one of the leading lights of the French government inquiry into UFOs. This was organized by the Space Centre at Toulouse and Dr Guérin, as an astrophysicist with the National Council for Scientific Research, spent much of his time in the 1970s looking at cases the authorities could not crack.

In a major paper in 1979, he summarized his years of research for *Flying Saucer Review* magazine (volume 25, number 1) and made some telling observations. He noted the three main contradictory stances to be found with ufology – that UFOs are either alien space probes, paraphysical manifestations, or social or psychological delusions. These are definitions that remain as true today as then, although with the addition of perhaps one new category, which sees UFOs as natural phenomena novel to science, capable of triggering altered states of consciousness and resultant hallucinations. This uncompromising battle-field saw each side refuse to concede ground to another. Guérin neatly called it a 'dialogue of the deaf'.

He rejects the classic model of alien space probes for many cogent reasons, notably the near certainty that no future advanced technology will make it as easy to cross the universe as the plethora of UFO sightings would otherwise suggest. Paraphysical theories go the same way, through lack of any evidence that UFOs can be conjured up as psychic manifestations and the fact that sightings tend to occur in spatial and temporal waves strongly linking them to certain places at certain times, rather than associating them with individuals in certain states of mind. Finally, Guérin dismisses the delusionary school of thought because of the statistical analysis done on unexplained cases by the team at Toulouse, which he feels has proved conclusively that UFOs are *real* phenomena. UFOs increase in proportion to the angle of view above the horizon, level of sunlight at a site, etc. Non-real things, for example, unprovoked hallucinations and delusions, would show no such correlation with visibility coordinates.

This is but a fraction of the lavishly funded work conducted over many years (and still continuing) which is probably the most detailed, scientific, and astonishing evidence that the UFO field has ever provided. Yet, despite the fact that they emerged via official French government projects, which were annually rebudgeted (and what government constantly rebudgets projects that are failures?), the published reports never made it into the scientific press: *New Scientist*, *Nature* or *Scientific American*, for example.

So what answer does Guérin propose that makes sense? His lengthy thesis reaches an extraordinary conclusion: 'modern ufonauts and the demons of past days are probably identical.'

Guérin talks of what he calls a magic technology with an intelligence that may stage events on a global scale in order to deliberately tease or mislead us. The purpose would be to manipulate social beliefs, establishing some level of circumstantial evidence for the visitation of aliens but not enough to persuade rational scientists. However, these 'demonstrations' would have an effect at the grass roots of society and could – given time – condition mankind into chosen patterns of thought and behaviour and into changing ideologies. Possibly, in this way, aliens could adapt humanity in a long-term and very subtle manner without direct intervention.

However, this is not the product of an alien culture from a nearby star, here to explore our solar system. Instead it is an intelligence which dwells in a different type of reality and moves through hyperspace in such a way as to interreact with us almost as a side-effect. It is an intelligence that is not here to study us but it has been here all along and effectively controls our whole existence! As Guérin phrased it: 'Is it possible then that – all unbeknown to us – we are a "colony"?'

This echoes the much-derided words of journalist and collector of weird tales, Charles Fort, who more than half a century ago speculated that 'we are property'. He suspected our mystification at what goes on is a result of the fact that we do not run the show as much as we think we do – if at all!

This monumental exercise in human thought based on a first-hand study of the UFO problem is all the more devastating when one appreciates that Dr Guérin is an astrophysicist, not some wide-eyed mystic. However, Dr Jacques Vallée, his fellow countryman (but long time resident of America) has a somewhat different approach.

Vallée, like Guérin, is an astrophysicist by training, but he is also a science-fiction author and now a leading computer wizard in California's Silicon Valley. He kindly supplied us with a copy of a paper he presented in June 1989 to the conference of the Society for Scientific Exploration at Boulder, Colorado. Its title is explicit: 'Five Arguments Against the Extraterrestrial Origin of Unidentified Flying Objects'.

Again, Vallée's thesis is extremely detailed. As the author of what were widely considered the first truly scientific UFO books, *Challenge to Science* and *Anatomy of a Phenomenon*, he has a pedigree unequalled in the field by any living scientist.

Here is a brief summary of Vallée's five arguments:

1. UFOs have been present throughout history and take on an image that is culturally relevant (from fiery shields in Roman and Greek times, unknown inventors' airships of the type in Jules Verne's novels at the turn of the century, to phantom Nazi weapons during the World War 2).

The Vehicle Assembly Building at the Space Center, where spacecraft are prepared for their missions. This single-storey building is so huge that clouds can form if precautions are not taken. (*Jenny Randles*)

This strongly suggests that the modern interpretation and perception of UFOs as extraterrestrial spacecraft is no more likely to be correct.

2. There are just too many bona-fide, unexplained UFOs for them to be space visitors. Vallée estimates some 14 million of them in just 40 years and points out that as our world is easily observed from space and we are beaming information freely to the cosmos by way of radio, TV and other communications, no aliens would need to waste such resources in this manner.

3. The aliens themselves, when seen, are also a major problem. Witness accounts 'indicate a genetic formulation that does not appear to differ from the human genome [the genetic make-up that uniquely describes a being] by more than a few percent.' These supposedly alien visitors 'not only resemble us but breathe our air and walk normally [in our gravity] ... [they] display recognizably human emotions such as puzzlement, interest or amusement.' For Vallée this is just not compatible with a genuine contact between humanity and real aliens.

4. The abduction cases offer a further serious problem, because in these stories aliens use their best technology when conducting almost obligatory, lengthy medical examinations on witnesses. Yet this technology is primitive and illustrates utterly wasteful and ludicrous medical procedures. While Vallée fails to note this, we might also add

that, even more significantly, the technology in abduction reports follows what ufologists call 'cultural tracking'. It is near the level of current technology, but just beyond it. We have made major scientific steps forward in the 40 years that these reports have been coming in, but when we compare how the alien technology has altered during this time, we see it is in tandem with human progress. It is never reported to be exceeding or diverging from it in any major sense.

5. Finally, Vallée notes that the cases possess clear signs of manipulation of space and time and witness consciousness, which better suit other theories.

In conclusion, Vallée says: 'Exciting as an extraterrestrial visitation to Earth would be, this paper has pointed out that in the current state of our knowledge UFO phenomena are not consistent with the common interpretation of this hypothesis.'

So what does Vallée find to be consistent? In his recent books, *Dimensions* and *Confrontations*, he talks of a 'control system' that regulates social behaviour and consciousness. He has likened it to a thermostat which alters the heat level in a house, raising or lowering it automatically by injecting heat or taking it away wherever necessary, so that a pre-set temperature is maintained.

In his paper he says that he is 'reserving judgement as to whether the control would turn out to be human, alien or simply natural' but considers two interesting possibilities: 'an alien intelligence, possibly earth-based, could be training us towards a new type of behaviour' or 'the human collective unconscious could be projecting ahead of itself the imagery which is necessary for our own long-term survival beyond the unprecedented crises of the twentieth century.'

As if these two divergent ideas were not fascinating enough, his third form of possible control mechanism seems fine-tuned to the 1990's way of thinking. He wonders if there might be a global system of control that is the cumulative result of all life in an ecological, physical, spiritual and consciousness sense, which seeks to keep our planet on a stable course and is presently compensating for the excesses mankind has wrought through the present technological age. As a result, it introduces UFO aliens as a visionary incentive to adapt our rather reckless attitude.

This certainly would explain the almost paranoid concern that oozes out of the messages from these aliens. In the 1950s they endlessly warned about a nuclear holocaust. In the 1960s they preached spiritual revolution. In the 1970s and 1980s we were warned about our disregard for life and pollution of the atmosphere. Now, with the mood of the day reflecting ozone-layer holes and global warming, we see a sort of cosmic

The gigantic Saturn rockets that were needed to lift man on his 'small step' to the moon. If aliens are visiting us, then it will require a totally different type of technology that we can only try to imagine. (*Jenny Randles*)

ecology developing within the subject. It is easy to predict how this will manifest in future alien contacts, as it is already doing.

Whatever the truth about UFOs, the fact that scientists are debating it at such a deep level and postulating quite remarkable and diverse theories from the evidence, surely justifies why the topic appears in this book. It also surely indicates to those critical and sceptical scientists that a balanced approach to the search for extraterrestrial life cannot do without some understanding of these issues, whatever you decide about them.

Perhaps scientists like Hynek and Vallée are correct in saying that the UFO phenomenon is more important than mere visitors from outer space. Possibly there is no connection at all. On the other hand, we must allow thinkers like Swords to continue to respond to those critics of the more straightforward extraterrestrial-visitation theories. Simply because they seem obvious and naïve, or appear to rely upon aliens who are little more than humans in spacesuits, does not mean that the viewpoints are incorrect. After all, if the aliens did engineer us and we are their colony or long-range experiment, we would doubtless find that they are like us – because in truth it is *we* who are like *them*. As the Bible phrases it: God has created man in his own image . . .

Finally, we have the esoteric ideas, such as Guérin's concept that aliens will not be found by flying a spaceship to another star but through quantum mechanics within hyperdimensional space, where 'reality' may be filled with countless strange intelligences. The aliens need not be light-years away, but cosmically just next door. Right now that sounds like complete nonsense, but who is to say that future research will not provide the key to unlock these other worlds, just as it recently gave us the key to unlock the atom?

On the other hand, as Hynek, Randles and science-fiction authors (notably Ian Watson in novels such as *Miracle Visitors* and the *Martian Inca*) suggest, maybe we do have alien visitors but they have never left their home world because of the physical constraints of the universe?

Perhaps the visions that we see are mere epi-phenomena, no more than our unconscious selves and our present culture desperately seeking a way to comprehend these alien intruders within our minds – psychic invaders from across the universe. If so, then are they probing for information, extracting data from us or taking genetic codes – a reversal of the 1960's science-fiction TV series *A For Andromeda?* In that story, earth scientists pick up genetic information via a radio communication from Andromeda; with it they are able to construct a living alien being.

Far off in an alien laboratory is some other intelligence already piecing one of us together? If UFO abduction cases with their genetic abstraction overtones are a psychological manifestation of real mental probings by aliens, this could be happening right now.

The Consequences

23
When the Aliens Arrive

In November 1985, when Mikhail Gorbachev brought his brave new world of Soviet glasnost to the meeting table with former enemy state America, something very curious happened.

President Ronald Reagan, in his final term, was eager to forge some sort of treaty on weapons reduction at this Geneva summit, hoping that it would pave the way for a new era of peace to replace the Cold War. Yet Reagan spoke in an unexpected manner after the conference. He suggested that if the earth were ever under attack from alien forces, America and the USSR would not hesitate to pool their technologies to fight a menace that would trample political boundaries in its threat to the entire planet!

Instantly, the UFO buffs picked up on these words. They saw clear proof that a huge cover-up was in force and that aliens were already threatening us. Of course, most people accepted that Reagan was using the Star Wars scenario to make the very sensible point that, fundamentally, the nations of the earth have much more in common than we realize. Reagan was noted for speaking his mind in terms that occasionally had his advisers grimacing. As a former film star, he may just have seen a few too many science-fiction films or was trying to express a good reason for global unity without thought for the way others would all too easily misinterpret his words.

Nevertheless, this comment by an American president does cast a new light on the theoretical questions reviewed in this book. For, if we take the majority view of science – which says that aliens are out there and in one form or another will eventually establish contact – then we must give serious thought to what all this implies. Will the world be a better or less stable place after the great event?

One advantage of a century of science-fiction literature, TV and films is that it has conditioned us almost to the point of normality about the topic. It now seems unlikely that mass panic would result if the news were ever broadcast to a stunned world. If contact came in the form of radio communications, the chances are the story would attract attention for a couple of days and soon be replaced by something considered more dramatic by the media moguls – for instance a TV soap star's latest romantic fling.

In a paper presented at the 37th Congress of the International Astronautical Federation in Innsbruck, Austria, in October 1986, Dr Allen Tough, a futurist at the University of Toronto, made this additional observation:

Millions of people already believe in extraterrestrials, yet continue working at their jobs, raising their children, and generally carrying on a normal life. Supermarket tabloids often announce extraterrestrial contact with large headlines, but no one panics. A recent Gallup Poll in the United States found that about half of all adults believe that intelligent life exists elsewhere in the Universe. Various movies and television series in the past few years have prepared people for positive contact with extraterrestrials.

Sociologist Dr Roberto Pinotti, who also spoke at the congress, thought that public exposure to such TV programmes was no accident. He cites the recommendations of astronomer Thornton Page, who was part of the American government's Scientific Advisory Panel on UFOs in the 1950s. Page suggested that UFOs be stripped of their aura of mystery, that a debunking programme be initiated and that civilian UFO investigative groups be infiltrated for 'subversive purposes'. Dr Pinotti concludes:

This document not only confirms that the negative impact of a possible ETI presence on Earth was carefully considered by US authorities in 1953, but also the fact that a definite educational programme was recommended and started since then.

Social and Psychological Viewpoints

A fascinating survey was conducted in 1988 by American researcher Ray Boeche, who asked people in mental-care professions (for example, psychiatrists) how they believed ordinary people would respond in a hypothetical, verified alien-contact situation. In that scenario, a TV message was to be released officially by the American government that admitted to a UFO crash at Roswell, New Mexico, in 1947 with alien bodies being recovered. Subsequent meetings with live aliens had occurred, such as one allegedly filmed at Holloman Air Force Base, and film of this would then be screened to an amazed world (paper to MUFON Conference, Lincoln, Nebraska, June 1988).

As you have seen, both these events are quite widely held to be true by some sections of the UFO community. For them, such a theoretical admission is by no means impossible in the very near future.

Boeche discovered that official research into this question had been carried out by several sources, including the CIA (in 1953) and the Brookings Institute for NASA (in 1961) when it was felt that 'intelligent

If we find intelligent life on another world, what will it look like? Perhaps we should ponder other intelligent life forms on our own planet. When contact is unequivocably confirmed, will it be with a bug-eyed monster, a humanoid or a dolphin? (*Jenny Randles*)

life might be discovered at any time . . . [and] the consequences of such a discovery are presently unpredictable.' Hence the need to find them out.

One of the most serious psychological problems was labelled 'future shock' by researcher Alvin Toffler. He suggested that the sudden confirmation of superior alien intelligences would lead to a traumatizing effect on society, causing it to give up its quest for knowledge or its investment of time, money and effort in finding resolutions to the problems facing the earth.

Of course, if future shock is the probable outcome, then that might be a reason for lack of confirmed alien contact. Advanced aliens may be well aware of the potentially devastating effects that overt communication would incite in a species and stay hidden.

In Boeche's survey of 475 professionals, 18 per cent responded and expressed their views. Their feelings were interesting and quite polarized, indicating a lack of true awareness of what would occur in such a situation. This may be justification for some (including the present authors), who feel that certain major UFO events may just possibly be sociological experiments, with official backing, that are seeded into the community to test the reaction of the public.

Boeche told us that his strongest result was that 65 per cent of respondents thought that alien contact would produce 'financial chaos'. Shattered religious beliefs, cultural allegiances and political aspirations were also widely considered to be under threat from his postulated situation.

Overall, Boeche advised us, the mental-health professionals likened the acceptance of an alien intelligence into our culture as 'equally as debilitating as when native American culture was effected by the colonisation of North America by white Europeans'.

Roberto Pinotto agreed with this conclusion in his address to the 1986 congress in Austria:

The malaise, mass neurosis, irrationality, and free-floating violence already apparent in contemporary life are merely a foretaste of what may lie ahead. Our culture is in constant turmoil, with its values incessantly changing and a dominating sense of general disorientation, involving also the weakest, least intelligent, and most irrational – and they are surely the majority members of society. The result is a cultural time bomb. And it could explode at any time. The news of the existence of ETI could prove devastating.

Interestingly, many of these problems were addressed in a remarkably responsible manner by the 1989 American TV series *Alien Nation* (developed from the film of the same title), which sought to show how we would cope with having to accept into our own society a whole refugee nation of basically humanoid but odd-looking extraterrestrials who are forced to come to earth. Notably, we rather arrogantly assume that human beings have the upper hand in this situation, whereas in fictionalized reversals, such as the TV series *The Invaders* or *V*, the aliens are regarded as hostile monsters out to murder, enslave and eat us!

Similar questions are looked at from another angle by Dr Allen Tough,

Dr Allen Tough is a futurist at the University of Toronto. He has lectured at many SETI gatherings on the alien question. (*Peter A. Hough*)

in his paper 'What do Extraterrestrials Plan for Our Future?' published in the *Journal of Transient Aerial Phenomena* in September 1987.

We discussed this idea with Dr Tough when he visited Britain, the topics ranging from overtly hostile alien intentions – about which he noted that despite the preference for hostility in films and books, they were the least supported by the evidence – through to the one already voiced which he felt was best supported: that some kind of subtle alien education programme might be afoot. However, while it was feasible that 'aliens might want to prepare for our future by gradually adding knowledge or influencing our beliefs' and one could read the evidence in this way, it should also be stressed that 'no persuasive evidence and certainly no proof of this' could be said to exist.

Earlier in the book we looked at some comments by film-producer Robert Emenegger who penned a 'fictional' reconstruction of aliens landing at Holloman Air Force Base. He did something quite interesting with this scenario: putting it to five leading social psychologists in America to see what reaction they felt it might provoke within society.

In fact, the results were confusing, with a wide range of opinions. A consensus suggested it may depend upon our prior beliefs. Those committed to UFOs would be euphoric, telling colleagues 'I knew it all along'. Whereas sceptics would be convinced it was a government hoax.

One counselling psychologist who has taken the subject much further is Dr Leo Sprinkle. He has worked with witnesses who believe they have had encounters with aliens and he stages annual get-togethers deep in the Rocky Mountains to 'see what will happen' when this extraordinary community of otherwise quite ordinary Americans gathers in one place. The charming name 'Ufolks' is used in place of 'aliens' and aptly expresses the homely feel of this happy band.

One thing that has emerged from these reunions, according to Dr Sprinkle, is the belief that there is 'a programme [that] can be called "cosmic consciousness conditioning" . . . The purpose of these encounters seems to be an initiation for the individual, and a stimulus to society, so that human development moves from Planetary Persons to Cosmic Citizens.'

This is done with 'games' that are played on humanity by alien forces in what almost amount to role-play exercises. In other words, the ludicrous nature of the information imparted (for example, 'I am Zookan from the Planet X on the other side of Saturn') is irrelevant in comparison with the cultural conditioning that occurs because of the tone of the messages overall.

Vox-pop Viewpoints

Vox pop – or the 'voice of the people' – is a device widely used by the media, which involves taking a random sample of opinions from passers-by on any current subject to gauge the mood in a country. It is interesting to test the method out on the question of the aftermath of alien contact, to see whether the Rocky Mountain ideology had filtered through to the grass-roots level. We have found that it appears to have done so. As authors of books on this subject, we receive huge volumes of mail from a broad cross-section of the public. Contact addresses are always offered as this sort of feedback is regarded as important in our work. This has produced some fascinating results.

Typical is 'Les' from Dundee, in Scotland, who writes to say that for 69 years he has followed science fiction. The moon landings deeply affected him by legitimizing 'dreamers' and making them a part of science. UFOs, for him, seem to be the product of an 'inner reality' representing outer truths.

'Evan' from North Carolina expresses the view that aliens are trying to communicate by imprinting crop circles into fields. They use new methods that we cannot fail to spot so as to warn of ecological disaster. Perhaps they are beaming thought power across the universe instead of radio messages and establishing contact through a method no scientist predicted.

On a similar theme, 'George' from Western Australia insists that the aliens are really angels and demons and that the earth is a cosmic battleground in a war between good and evil. What we might regard as proof of alien contact is in fact a side-effect washing over society, depicting our pawn-like nature in a cataclysmic struggle between the powers of darkness and light.

The same theme is taken up by 'Jim' from South Wales, who writes in great detail about 'the divine plan' that sees the earth as a living force, itself in rebellion with all the dire ecological consequences that we face. 'Our friends in space' are said to be responsible, making these visible signs (everything from earthquakes and climatic changes to UFOs and crop circles) to help to steer us towards a universal goal of purity.

Over in Germany, one young woman was reporting strange dreams that prompted a major change in her lifestyle. She reported regular, powerful experiences with a very beautiful woman with white skin and vivid blue eyes who kept showing her scenes of the earth filled with pollution and heaping the blame quite specifically on the dreamer herself. She awoke more ecologically aware, feeling that she must do something about these issues.

All of these communications came within a few weeks in February and March 1990 and they reflect quite well the kind of material that flows in at a prodigious rate to us all of the time. There is little doubt that a cultural shock wave is reverberating around the planet which, as yet, has no clearly defined area of research to bring it all together.

What if aliens are communicating at a level of planetary consciousness and endeavouring to change society from within? Would we not see thousands of 'atuned' people absorbing this information, changing their views and their lifestyles and evaluating it in a personally relevant form? It is a bizarre possibility, but one that cannot be dismissed out of hand.

The Dream Makers

Some say that our outward quest for alien contact is just a sham, because these messages are already coming in and conditioning society in subtle ways. Some ufologists have even argued that films like Spielberg's *Close Encounters* and *ET* were part of this conditioning process. He was 'inspired' by the aliens to produce them and change the feelings of many millions about extraterrestrial life. There is a nice romantic element in that idea, but it lacks any substance.

On the other hand, science-fiction authors like Ian Watson recognize the importance of consciousness and cultural conditioning in their work. Watson has written several trail-blazing novels which explore the outer limits of the mind and the way that it might make a far better instrument of long-range communication than rapidly outmoded technology.

Watson's work is shot through with imagery of special states of consciousness, aliens who dream, and our need to turn within for answers as opposed to constantly looking outward. His seminal work about UFOs, *Miracle Visitors*, even developed the idea of 'UFO consciousness' – a state of mind in which the earth itself sings to us and introduces 'unidentifieds' into our midst, where our UFOs come to life via the catalytic effect of psychically aware witnesses.

Apparently unknown to Watson, this has become a major theme of UFO research as the evidence often seems to be pointing in that direction. Indeed he reports that, as he wrote this deep and perceptive novel, UFOs homed in on Oxfordshire where he lived, as if emphasizing the very symbiosis that he spoke of.

Author David Langford discussed with us the potential impact of films such as *Close Encounters*, terming it 'pernicious', by which he meant that our culture tends to adopt what he terms the mood of 'cargo-cult philosophy; lie back, relax, don't think and the nice aliens will come and give you transcendence right at home.'

There is probably much truth in this, as we have become a passive almost receptive society. But are we waiting for the aliens rather than actively seeking them because we have grown weary of science and prefer mysticism? Or is it because we are subliminally aware that we should listen out for the real messages now coalescing inside ourselves?

Perhaps alarm bells are ringing within the buried hallways of our consciousness and some of us are already stirring in response. Have the major cultural changes – such as the sudden stride towards freedom, democracy, anti-nuclear treaties and peace, the vast and expensive plans to solve ecological problems, or the incredible responses to campaigns to beat world hunger, mobilizing whole sections of our society – been mere accidents of history or are they a result of those conditioning messages that are emerging within?

Langford notes that after *Close Encounters*, witnesses to UFO abductions reported the same type of child-like aliens found in the plot, whereas nobody ever reports huge black monoliths, like the one in *2001: A Space Odyssey* – a film which had a similarly profound emotional effect one generation earlier. For many this is evidence of how films shape experience, but the relationship is really much more complex than that.

Spielberg based his aliens (albeit exaggerating them in form) on genuine reports, as he did with many of the scenes in the film. So which came first? Did Life imitate Art or Art imitate Life, or is it both and neither – are we dealing with an interface where fantasy and reality become meaningless terms, a place where an alien consciousness resides, as George Harrison said, 'within you and without you'.

Writer Paul Barnett (alias John Grant) believes that many will respond to the ultimate proof of alien contact by going out on to the streets and celebrating. It 'could give humanity the biggest cultural boost since the human animal came into existence . . . [and] would inject a much needed dose of humility into our species, making us re-examine a lot of our basic tenets.'

He feels that one of the best consequences will be the destruction of political demagogues. All over the world there are rulers who believe it is their right to impose their will on others. An uneasy acceptance that there are higher (or at least more intellectually responsible) beings ought to bring empires crashing down around our feet.

In a way, this is just what we see happening at the start of the 1990s. Again, is it coincidence, or a result of a growing social acceptance that we are not alone and may be inwardly convinced of that fact already?

This mood of near complacency provides great support for those who say acceptance of alien visitors would be easy.

In February 1990 Jenny met John Leeson, the actor behind K-9, the famous cyborg (part machine, part flesh and blood) dog in the perennial TV favourite *Dr Who*. They met during the filming of a TV programme at the BSB studios in London.

Leeson noted in conversations between takes that for a long time he was quite ambivalent about the question of alien life, but that 'with all those planets out there in our solar system and beyond I think there has to be life . . . Fighting Cybermen was just a job of work at the time, but now I think there maybe has to be more to it. Of course, if it happened and it arrived on earth then the aliens would just be accepted. There would be no upheavals because we would have to believe.'

Professor Sir Francis Graham-Smith holds similar views, based on the premise of a confirmed extraterrestrial signal from across the galaxy. 'I don't think it would make any difference to the way we go about day-to-day living. I'm sure we would just carry on making cups of tea and so on. It would be stimulating but not revolutionary.'

One man who ponders this question from a number of different points of view if Patrick Tilley. Not only has he had some interesting experiences himself which satisfy him that there are hidden depths to the mind, but he has worked on both the film- and novel-writing sides of the alien dream market.

Tilley was the man who wrote the film *The People that Time Forgot*, with its weird catalogue of monsters. He left the cinema to concentrate on novel writing because he felt that the commercial constraints on films often interfered with the creative expression: it was less a case of saying what made sense or what you inwardly felt was likely to happen and more one of what the budget or the wardrobe department dictated. He told us: 'A lot of the SF movies are horror films just using another genre.' After quitting films, he produced what many believe to be a real masterpiece of fictional drama about contact between man and aliens.

Entitled *Fade Out*, because the story begins with a sudden and dramatic fade out of communications all over the earth, it delves deeper into the interrelationship than any film could. Indeed, the one planned for this superb story, as we have already mentioned, was unfortunately sidelined in the shadows of Spielberg's *Close Encounters*. Yet it would have made a much more intellectually stimulating film. Tilley says of his novel:

I commented on how our ideas about aliens are shaped by 'fifties movies. And I posed what to me is a critical question: just how would we recognize an alien intelligence? We might argue that an alien needs an opposable thumb in order to build spacecraft with tools, but that is a presumption. They might use the mind to structure crystals.

The aliens seem as intangible as ever.

A revised edition of his novel, first published over a decade ago, appeared in 1989. In this Tilley said:

I was able to put back material that was omitted because the original editor asked for it to be cut. In one scene an alien probe brings our own Voyager plaque back to earth with all the symbols and messages on it that we think will make sense to another culture. I ask whether the return of the plaque was the purpose of the visit or were the aliens maybe bringing back our visiting card that we have ejected into space? Possibly it was doing the universe a favour by making sure that nobody else knows we are here!

This makes one think of the movie *The Gods Must Be Crazy*, where a primitive tribesman discovers an artefact from a god-like species of 'higher intelligence'. It has been dumped on earth and in the true spirit of researchers like Erich von Dänekin, he sets out on a long quest to find the vastly superior beings who have sent this sign. The problem is that the alien artefact is nothing more than a fizzy drink's container discarded from an overflying aircraft by an intelligence which in many respects was considerably inferior to the hero of the film!

This reverence that we tend to impose on anything unknown leads us towards new questions of great importance, because as Patrick Tilley said, 'I would like to think that alien life is out there. But if it is, then do they worship a god or embrace that sort of concept?'

The dying extraterrestrial in Greg Bear's brilliant novel, *The Forge of God*, is asked by a future President Crockerman, whether it believes in God. The alien replies: 'I believe in punishment.'

Religious Perspectives

Dr Barry H. Downing, author of *The Bible and Flying Saucers*, is a Presbyterian pastor. He conducted a very interesting survey for MUFON in 1988. He asked 100 Protestant and Roman Catholic ministers how they believed religion would be affected by the news that aliens did exist and were coming to earth. Their views split pretty well into two groups.

A minister in one group felt that there would be little or no change, citing that God could be said to hold domain over other planets and – in one Roman Catholic example – that the aliens would soon 'adopt our sinful ways and would learn of God and sin and redemption'.

The other group saw the news as a challenge that would positively expand theological knowledge, allowing them to ask 'Who is your God?' and 'Are you a fallen race with a saviour such as Christ?'

Interestingly, there was little if any feeling among the ministry that alien contact would be destructive towards religion, even though this was seen as one of the main negative influences by other parts of society.

Perhaps religion is merely over-confident about its own ability to re-evaluate itself in the face of alien contact. How, for example, would a truly alien life form fit the Biblical concept that God created man in His image? This does seem a supremely egotistical concept. Would it shake belief if a race of giant slugs were equally convinced that God 'really' looked like a giant slug?

Probably the most powerful way in which alien contact affects religious thinking is how it becomes tied up with the apocalypse.

According to the Bible, prior to the second coming of Christ (considered imminent by many scholars), there will be the battle of Armageddon between the angels and the demons. The highly coloured and symbolic Book of Revelations in the New Testament is full of images of destructive earthquakes, plagues that wipe out millions (said by some to be AIDS) and 'signs in the sky' as warnings that the 'End Times' are here.

While many Christian fundamentalists take this very seriously, even if allegorical of mankind's need to change his wicked ways, some regard it as fact. Whole religious communities fervently believe that only 144,000 people will survive the holocaust and that these chosen ones will be airlifted to safety by what we might consider to be alien beings.

The Saints of God Church International is a good example. They regularly mailshoot the sort of literature most people never read but file in the nearest waste-bin. As authors of 'heretical' books which look at the questions of alien intelligence from the fallen realms of science, we receive it by the cart-load!

The Church is adamant that the 1990s is the 'final decade' and that 'The True Aliens' (as one pamphlet is titled) are not 'little green men from Mars . . . [but] actually much more close to home'. For them we are in the early stages of 'Operation Earth' in which 'the Interplanetary Council of the Sovereign God has deemed it necessary for the salvation of planet earth to launch a MASS INVASION . . .' This will take the form of destructive attacks against the 'Antichrist' and the 'Madhi' who rule the two key power blocks on earth – essentially the West and the East. Jesus is said to be the leader of this invasion fleet of aliens, already moored off earth and watching closely as we are given our last chance via constant warnings to behave – or else!

We may smile at these beliefs, but they have a grip over large sections of humanity. It is surprising (or maybe not so surprising given its implications for them personally) that in Dr Downing's survey, the Church itself was almost seen to be ignoring this interpretation of alien contact.

Whole magazines exist which discuss the kind of experience which ufologists would term 'alien abductions' in wholly different forms. The March-April 1990 issue of *The Four*, subtitled 'A Christian Newsletter of the Tribulation', features the report of a case from Kansas in November 1980 which developed to the point where the entity was manifesting in front of one woman 'as a being of light, or "Starman" . . . [who] proved that he was not of this world . . . [but] the original Peter the Apostle!'

Ironically, during that same month in Todmorden, north England, a police officer had a visionary experience with a Biblical figure in a white robe and beard, who described himself as 'Yosef'. Yet this encounter is listed as one of the key cases in UFO literature of evidence for some sort of alien contact. Clearly, only the cultural contexts are different and dictate how these experiences become incorporated into our thinking.

As David Barclay phrased it in an article for the magazine *UFO Brigantia* (January 1990) are 'Flying Saucers – Harbingers of Doom?'

Probably there is no fundamental truth to all of this, merely variations on a theme. What it may say above all else is that we mythologize as godly anything that seems to be beyond ourselves. The idea of contact with alien powers, which is now becoming acceptable to our society, is presented in a way that the individual can assimilate. If this means junking orthodox religion for a neo-religion based on 'the extraterrestrials', then so be it. If it means moulding standard religion so that its gods literally become aliens and so offer proof that 'they' are coming, or were here long ago, then fine. If, instead, it requires reshaping the experiences or beliefs about aliens entirely so that they become messengers of God, then that is also a way of coping.

As to whether any of these interpretations are 'true' beyond a subjective level, perhaps time alone will tell.

The Scientific Dilemma

NASA is hopeful that its latest searches for life in the universe may provide the ultimate proof. It has even set up a code of practice.

Released documents, dated 29 December 1988, are titled 'SETI Post Detection Protocol'. They refer to the SETI Microwave Observing Project, just getting under way, which will scour the cosmos for signals in a narrow range of frequencies during an intense probe. This will be '10 billion times more comprehensive than the sum of all previous searches.' No wonder they decided to draw up a plan of action for what to do when the aliens are contacted!

The document admits that 'there is at this time no policy agreed to by the governments of the world for the dissemination of information about a verifiable ETI [extraterrestrial intelligence] signal.' However, 'an

The Multi-Channel Spectrum Analyser, version 2. The chip used in this equipment was designed at Stanford University. It replaces the equivalent of three boards in the previous generation spectrum analyser, MCSA 1. Technology like this is on the front line of NASA's search for extraterrestrial intelligence. (*NASA*)

emerging consensus ... [says] the detection of an extraterrestrial civilization is a discovery with such profound implications that it transcends national boundaries and should be the property of all humankind.'

Nevertheless, the list of procedures to be carried out after detection of a signal are extraordinary. They involve checks to rule out interference, malfunctions, hoaxing, and terrestrial spacecraft. They then require continuous reception by other sources, replacement of all software in the system – plus other steps before NASA headquarters is even informed.

Only then would a meeting of appropriate experts be convened to decide upon the status of the message and, after much else, 'at the discretion of the Administrator of NASA, a formal announcement by him or the President or both may be made and broadcast.' However, the full scientific results 'will be published as soon as possible in the open literature.'

The reason for all of these precautions is the understandable one of not raising the alarm by 'crying wolf' with claims that later prove to be unjustified. So far, it is alleged, all stories of unidentified signals have been false alarms because they have never been duplicated or proven to be intelligent in origin.

This mood of optimism is certainly filtering through to scientists. We spoke briefly to Professor Heinz Wolff, the man in charge of Britain's own project which put its first astronaut into space as a crew member on the Soviet flight, code-named Juno. He was delightfully speculative that life will be found among the millions of stars and was surprisingly refreshing in his appreciation of the less straightforward categories of evidence. 'I have met a man who claims to have had contact with aliens,' he told Jenny. 'While I cannot say what the truth is, I think he was sincere.'

Meanwhile, Dr Paul Davies has transferred to Australia, where the NASA search of the southern hemisphere will open up many new stars for potential exploration for the first time.

'I cannot myself think of a more profound discovery,' he said. 'I have often been asked what would be the best – or at least the most exciting – new development of the 'nineties and it must be contact with ETI. Its effects would be undoubtedly sensational. I can imagine no scientific discovery that would be more important.'

He added that scientific revolutions do not always impinge directly on the lives of the ordinary person, but they can still profoundly affect them in ways they do not notice:

You only have to look at the power of religion in our so-called scientific age. It forges world views. With the discovery of ETI we are talking about something of comparative emotional magnitude. It would change how people looked at the meaning of their lives in ways that we cannot even predict. Indeed, I hope that the discovery of ETI would be like an enhanced version of the mood of global unification that the moon shots provided. That it might make us see ourselves as more of one planet and less a collection of separate societies. If ETI was found, the world, quite simply, would never be the same again.

American astronaut and scientist Dr Brian O'Leary was even more philosophical in a statement he made in 1989 as he publicly launched his outspoken views on alien contact. He warned that now was the time for radical reform, with our international relationships currently on such an upturn. We required a new reality to replace the old and 'part of that new reality is embracing the unknown and the possibility that we have visitors beyond Earth or beyond our dimensions of time and space.'

Echoing the mood of philosopher Joseph Campbell, he added words that form a fitting conclusion to this survey. They form a battle-cry for science, non-science and humanity.

'It is time we let go of our fears and evolve into a higher awareness of ourselves and our place in the universe.'

Afterword

What was presented as a momentous event occurred as we were finishing this book. Jodrell Bank announced to the world on 24 July 1991 that they had discovered the first confirmed new planet outside our solar system. Instantly, this made very plausible the idea that other worlds were commonplace and that alien life was a reality. It came at a vital time for the radio telescope installation, which was being squeezed by severe government funding cuts. However, there is no reason to think this was anything but a *bona fide* scientific discovery.

As Professor Sir Graham-Smith explained in his interview with us, Jodrell Bank is not an optical observatory so the planet was never 'seen', as some media sources misunderstood. In fact, it was detected by chance through a study of radio waves emitted by a pulsar 30,000 light years away.

The discovery team headed by Professor Andrew Lyne was several years into a project studying the neutron star at the heart of a huge energy powerhouse when post-graduate researcher, Setnam Shemar, found an odd signal pattern which indicated that the star's emissions were varying. Eventually it was concluded that this seemed consistent with a planet interfering with the parent star's motion through space.

This triggered a brief media ballyhoo. Banner headlines such as: 'Out of This World', proclaimed the event in historic proportions. One source contended that Jodrell Bank had effectively pronounced that 'flying saucers and little green men' were now possibly not 'just the product of an over-active imagination'.

Strangely, some reports featured the claim that this was the first-ever indication of other planetary systems. This is not the case as several planets much closer to our own solar system have been deduced using different but not dissimilar methods. Initial indications are that the body orbiting pulsar 'PSR 1829–10' is a lot more massive than Earth and, given the awesome nature of its stellar parent, very unlikely to have given birth to any life at all.

After a honeymoon period of several days, the story died a quick death. Indeed, during the summer of 1991, crop circles received twenty times as much press coverage as Jodrell Bank's discovery.

The crop circles mystery itself seemed to fall apart in September, with the revelation that two practical jokers, Doug Bower and Dave Chorley, had been hoaxing circle configirations for the last thirteen years. Their hoaxes included many of the most complex designs found in southern England.

The aliens seem as intangible as ever.

Further Reading

CHAPTER 1 Who Goes There?

Science Fiction Monthly. Volume 1, number 11, 1974.
Encyclopedia of Science Fiction. Octopus 1978.
Brian Aldiss, with David Wingrove: *Trillion Year Spree*. Paladin 1988.

CHAPTER 2 Aliens – Haunters of the Dark

Jack Sullivan: *The Encyclopedia of Horror and the Supernatural*. Viking Penguin 1986.
Brian Aldiss, with David Wingrove: *Trillion Year Spree*. Paladin 1988.
H.P. Lovecraft: *Omnibus 3: The Haunter of The Dark*. Grafton 1988.
Stephen King: *The Tommy-Knockers*. Hodder & Stoughton 1988.
Ramsey Campbell: *Midnight Sun*. Macdonald 1990.
Ramsey Campbell: 'The Voice on the Beach' (short story) in *Fantasy Tales*, volume 5, number 10.

CHAPTER 3 Alien Dreamer

Bob Shaw: 'An UnComic Book Story' in *Science Fiction Monthly*, volume 2, number 9.
Bob Shaw: *The Ragged Astronauts*. Gollancz 1987.
Bob Shaw: *The Wooden Spaceships*. Gollancz 1988.

CHAPTER 4 Alien Thinking

Ian Watson: *The Embedding*. Gollancz 1973.
John Grant: *A Directory of Discarded Ideas*. Ashgrove 1981.
John Grant: *Dreamers*. Ashgrove 1984.
John Grant: *Sex Secrets of Ancient Atlantis*. Grafton 1985.
John Grant: *Great Mysteries*. Chartwell 1989.

CHAPTER 5 William Loosley – Documentary Proof?

David Langford: *An Account of a Meeting With Denizens of Another World*. UK: David & Charles 1979; USA: St Martin's Press 1980.
Nigel Blundell and Roger Boar: *The World's Greatest UFO Mysteries*. Octopus 1983; reprinted 1991.
David Langford and John Grant: *Earthdoom*. Grafton 1987.
Colin Andrews and Pat Delgado: *Circular Evidence*. Bloomsbury 1989.
Terence Meaden: *The Circle Effects and its Mysteries*. Artetech 1989.
Jenny Randles and Paul Fuller: *Crop Circles: A Mystery Solved*. Hale 1990.
Whitley Strieber: *Majestic*. Macdonald 1990.

CHAPTER 6 Aliens on Film

J. Allen Hynek: *The UFO Experience*. Abelard-Schuman 1972.
Robert Emenegger: *UFOs Past, Present and Future*. Ballantine 1974.
Steven Spielberg: *Close Encounters of the Third Kind*. Dell 1977.
Jenny Randles: *From Out of the Blue*. Inner Light 1991.
Jenny Randles and Paul Whetnall: *Alien Contact*. Coronet 1983.
Jenny Randles, Brenda Butler and Dot Street: *Sky Crash*. Spearman 1984; updated, Grafton 1986.
John Fairley and Simon Welfare: *Arthur C. Clarke's Chronicles of the Strange and Mysterious*. Grafton 1987.
Patrick Tilley: *Fade Out*. Sphere 1989.
Jacques Vallée: *Confrontations*. UK: Souvenir; USA: Ballantine 1990.

CHAPTER 7 Have Spaceships Landed?

Carl Jung: *Flying Saucers: A Modern Myth of Things Seen in the Sky*. Routledge & Kegan Paul 1959.
Jerome Clark and Loren Coleman: *The Unidentified*. Warner 1975.
Allan Hendry: *The UFO Handbook*. Sphere 1980.
Jenny Randles: *Abduction*. UK: Hale 1988; USA: as *Alien Abductions*. Inner Light 1989.
Philip Klass: *UFO Abductions: A Dangerous Game*. Prometheus 1989.
John Spencer: *Perspectives*. Macdonald 1990.
Jerome Clarke: *Encyclopedia of UFOs* (3 volumes). Omnigraphics Inc., Detroit 1990–92.

CHAPTER 8 George Adamski – First Contact

Desmond Leslie and George Adamski: *Flying Saucers Have Landed*. Werner Laurie 1953.
George Adamski: *Inside the Spaceships*. Spearman 1956.
Lou Zinsstag and Timothy Good: *George Adamski: The Untold Story*. CETI 1983.
The Probe Report. April 1983.
Leonard Cramp: *Space, Gravity and the Flying Saucer*. Werner Laurie 1954.
Gray Barker: *Book of Adamski*. Saucerian Publications 1965.

CHAPTER 9 Life on Mars

The Unknown. December 1986 and January 1987.
Magonia. July 1986, number 23.
UFO Universe. January 1990.
Cedric Allingham: *Flying Saucer from Mars*. Muller 1954.
Robert Chapman: *UFO – Flying Saucers Over Britain*. Granada 1969.
Peter Haining: *The Race for Mars*. COMET 1986.
Brian O'Leary: *Exploring Inner and Outer Space*. North Atlantic Press 1989.
Richard Hoagland: *The Monuments of Mars*. North Atlantic Press 1987.

CHAPTER 10 Channelling and the Inter-Space Connection

Brad and Francis Steiger: *The Star People*. Berkley 1981.
Shirley MacLaine: *Out on a Limb*. Bantam 1983, Corgi 1984.
Ruth Montgomery: *Aliens Among Us*. Fawcett Crest 1985.
Diane Tessman: *The Transformation*. Inner Light 1988.

CHAPTER 11 Inside the Aetherians

George King: *You Are Responsible*. Aetherius Publications.
Richard Cavendish (editor): *Man, Myth and Magic*. Purnell 1970 part-work.

CHAPTER 12 Somewhere over the Interstellar Rainbow

Walter Sullivan: *We Are Not Alone*. Hodder & Stoughton 1965.
David Holmes: *The Search for Life on Other Worlds*. Bantam 1966.
Fred Hoyle and Chandra Wickramasinghe: *Life Cloud*. Dent 1978.
Ian Ridpath: *Messages from the Stars*. Fontana 1978.
Stuart Holroyd: *Alien Intelligence*. David & Charles 1979.

CHAPTER 13 Eye on the Sky

How We Know About the Universe. Jodrell Bank publication.
Jill Tarter: *Radio Frequency Interference at Jodrell Bank Observatory Within the
    Protected 21cm Band*. Pergamon 1989.

CHAPTER 14 The Search for Other Worlds

Paul Davies: *Other Worlds*. Dent 1980.
Paul Davies: *God and the New Physics*. Penguin 1983.
Paul Davies: *The Cosmic Blueprint*. Heinemann 1989.

CHAPTER 15 Signals from Space?

Duncan Lunan: *Man and the Stars*. Souvenir 1974.
Anthony Lawtson and Jack Stoneley: *Is Anyone Out There?* W.H. Allen 1975.

CHAPTER 16 Was God an Astronaut?

Erich von Dänekin: *Chariots of the Gods?* Souvenir 1969.
Erich von Dänekin: *In Search of Ancient Gods*. Corgi 1976.
Erich von Dänekin: *Gold of the Gods*. Corgi 1975.
Reñe Noorbergen: *Secrets of the Lost Races*. New English Library 1977.
Ronald Story: *The Space Gods Revealed*. New English Library 1977.
Robert K.G. Temple: *The Sirius Mystery*. Futura 1976.
The Bible: revised standard version.
Edward Ashpole: *The Search for Extraterrestrial Intelligence*. Blandford 1989.
The Unknown. August, September, October, November 1985.
Donald Keyhoe: *Flying Saucers: Top Secret*. G.P. Putnam's Sons 1960.
Josef Blumrich: *The Spaceships of Ezekiel*. Bantam 1974.

Clifford Wilson: *Crash Go the Chariots*. Lancer Books Inc. 1972.
Clifford Wilson: *UFOs and Their Mission Impossible*. Signet 1974.

CHAPTER 17 Aliens in Orbit

Don Wilson: *Our Mysterious Spaceship Moon*. Dell 1975.
Maurice Chatelain: *Our Ancestors Came From Outer Space*. Pan 1980.
Timothy Good: *Above Top Secret*. Sidgwick & Jackson 1987.

CHAPTER 18 Crashed Spaceships and Dead Aliens

Quest UFO. 1977 issue.
Official UFO. November 1976.
True Flying Saucers & UFOs Quarterly.
Charles Berlitz and William Moore: *The Roswell Incident*. Granada 1980.
Jenny Randles: *The UFO Conspiracy*. Blandford 1987.
International UFO Reporter. September and October 1987.
UFOs Canada – A Global Perspective. 13th Annual MUFON UFO Symposium
    Proceedings 1982.
UFOs – The Burden of Proof. MUFON UFO Symposium Proceedings 1985.
John Baxter and Thomas Atkins: *The Fire Came By*. Doubleday 1976.
Jenny Randles: *Abduction*. UK: Robert Hale 1988; USA: Inner Light 1989.

CHAPTER 19 Aliens in Focus

Charles Berlitz and William Moore: *The Roswell Incident*. Grafton 1980.
Jenny Randles: *The UFO Conspiracy*. Blandford 1987.

CHAPTER 20 Alien Properties

Flying Saucer Review. Volume 22, number 4 1976.
Jenny Randles and Peter Hough: *Death by Supernatural Causes?* Grafton 1988.
John G. Fuller: *The Interrupted Journey*. Souvenir 1980.
Jenny Randles: *Pennine UFO Mystery*. Granada 1983.

CHAPTER 21 Back in the USSR

James Oberg: *UFOs and Outer Space Mysteries*. Donning 1982.

CHAPTER 22 Where Are They?

Jacques Vallée: *Challenge to Science*. Spearman 1966.
Jerome Pearson: 'Lonely Life of a double planet', *New Scientist*, August 25 1988.
Jacques and Janine Vallée: *Anatomy of a Phenomenon*. Spearman 1967.
Brian Aldiss: *Helliconia Spring*. Gollancz 1982.
Michael Swords: 'The Third Option', in *International UFO Reporter*, Jan/Feb
    1987.
Jacques Vallée: *Dimensions*. Souvenir 1988.
Jacques Vallée: *Confrontations*. Souvenir 1990.

Pierre Guérin: 'Thirty Years after Kenneth Arnold', *Flying Saucer Review*, Volume 25, number 1.

CHAPTER 23 When the Aliens Arrive

Barry H. Downing: *The Bible and Flying Saucers*. Lippincott 1968.
Ian Watson: *Miracle Visitors*. Gollancz 1978.
Brian O'Leary: *Exploring Inner and Outer Space*. North Atlantic Books 1989.
Dr Allen Tough: 'What Do Extraterrestrials Plan for Our Future?' in the *Journal of Transient Aerial Phenomenon*, September 1987. Dr Allen Tough: *'A Critical Examination of Factors that might Encourage Secrecy'*. Dr Roberto Pinotti: *'Contact: Releasing the News'*. Both papers presented to the 37th Congress of the International Astronautical Federation, Innsbruck, Austria, in October 1986.
Ray Boeche: *'Public reaction to alien contact: a study'*. A sociological paper presented to MUFON Conference in Nebraska in 1988.
Greg Bear: *The Forge of God*. Gollancz 1986.

Magazines referred to in the text

FATE, P.O. Box 64383, St Paul, Minnesota 55164–0383 USA
Flying Saucer Review, FSR Publications Ltd, Snodland, Kent, ME6 5HJ, England
Fortean Times, 96 Mansfield Road, London NW3 2HX, England
International UFO Reporter, 2457 West Peterson Avenue, Chicago, Illinois 60659, USA
Journal of the British Astronomical Society, Burlington House, Piccadilly, London, England
Journal of the British Interplanetary Society, 27–29 South Lambeth Road, London, SW8 1SZ, England
Magonia, John Dee Cottage, 5 James Terrace, Mortlake Churchyard, London SW14 8HB, England
MUFON Journal, 103 Oldtowne Road, Seguin, Texas 78159–4099, USA
New Scientist, Commonwealth House, New Oxford Street, London WC1A 1NG, England
Omni, Omni Publications International Ltd, 909 Third Avenue, New York, NY 10023, USA
PROBE, c/o BUFORA, 103 Hove Avenue, Walthamstow, London E17 7NG, England
Skeptical Enquirer, 560 'N' Street South West, Washington D.C. 20024, USA
UFO Brigantia, 84 Elland Road, Brighouse, West Yorkshire, HD6 2QR, England
UFO Times, c/o Mike Wootten, BUFORA, Suite 1, The Ley, 2c Leyton Road, Harpenden, AL5 2SP, England

UFO Call (UK only), phone 0898 121886

The authors may be contacted directly by writing to 37 Heathbank Road, Stockport, Cheshire SK3 0UP, England

Index

Page numbers in *italic* refer to illustrations.